AF709325

Environment Policies for the 1980 s

ORGANISATION
FOR ECONOMIC
CO-OPERATION
AND DEVELOPMENT
PARIS 1980

The Organisation for Economic Co-operation and Development (OECD) was set up under a Convention signed in Paris on 14th December 1960, which provides that the OECD shall promote policies designed:

— to achieve the highest sustainable economic growth and employment and a rising standard of living in Member countries, while maintaining financial stability, and thus to contribute to the development of the world economy;
— to contribute to sound economic expansion in Member as well as non-member countries in the process of economic development;
— to contribute to the expansion of world trade on a multilateral, non-discriminatory basis in accordance with international obligations.

The Members of OECD are Australia, Austria, Belgium, Canada, Denmark, Finland, France, the Federal Republic of Germany, Greece, Iceland, Ireland, Italy, Japan, Luxembourg, the Netherlands, New Zealand, Norway, Portugal, Spain, Sweden, Switzerland, Turkey, the United Kingdom and the United States.

***

© OECD, 1980
Queries concerning permissions or translation rights should be addressed to:
Director of Information, OECD
2, rue André-Pascal, 75775 PARIS CEDEX 16, France.

363.705
Or3e

FOREWORD

This publication includes the papers prepared for the Meeting of Ministers of Environment of OECD Member countries held in Paris on 7th and 8th May, 1979.

The OECD environmental programme is now ten years old, and this was the second meeting of OECD Ministers of Environment. The Ministers discussed changes in the state of the environment in OECD Member countries that have occurred during the past decade of environmental action. A report on this is available in a separate publication. The Ministers also examined the prospects for the future.

The outcome of the meeting and the views of the Ministers are reflected in the Communiqué and in the Declaration. These are presented in the first paper.

The other four papers formed the background to the Ministerial discussions. Three of the papers were prepared by the Secretariat of the OECD assisted by a number of prominent, widely recognised experts in the fields of environment and economics, including Nobel Prize Winner Professor Leontief. The fourth paper was prepared by J. F. Saglio, former Délégué à la Qualité de la Vie in the Ministry of Environment and Quality of Life, Paris.

CONTENTS

I

MINISTERIAL STATEMENT AND DECLARATION

1. Assessing the action undertaken over the past few years, Ministers for Environment of the OECD countries laid the foundations of the policies they proposed to implement during the coming decade and decided to strengthen their co-operation in a number of vital fields. On 7th and 8th May, 1979, at the Château de la Muette, Paris, they held a meeting chaired by Mr. Douglas M. Costle, U.S. Environmental Protection Agency Administrator. Vice-Chairmen were Mrs. Ingrid Leodolter, Minister for Health and Environment (Austria), Mr. Günter Hartkopf, Secretary of State, Ministry of the Interior (Germany), Mr. Senichiro Uemura, Minister of State, Director General of the Environment Agency (Japan), Mr. Eric Enlund, Minister of Agriculture responsible for environment policy (Sweden).

2. Ministers had before them the first international report on the State of the Environment in OECD Member countries.[1] On the basis of their national experience they confirmed progress made over the past decade in a number of fields. They highlighted in particular, achievements concerning the control of pollution in fresh water bodies and air pollution in urban areas; the reduction of the flow into the environment of certain persistent chemicals; the extension of national and regional parks; and the protection and rehabilitation of sites of social or cultural value.

3. Ministers pointed out that these improvements in environmental quality had generated significant benefits, without in most cases negative effects on the economy. According to the Secretariat reports before the Ministers, the short-term net employment effects have been positive; the impact on inflation has been on the whole moderate, averaging 0.1 - 0.3 per cent per annum; the impact of environmental policies on the rate of economic growth in the short run has been neutral, perhaps even slightly positive. Overall, expenditures by OECD countries on pollution abatement amounted to 1 to 2 per cent of GNP while certain studies indicate that damage from pollution has ranged between 3 and 5 per cent of GNP. Specific benefits have included reduced morbidity and mortality, improved productivity in land and labour, technological innovation and increased amenities. Ministers also noted that in some

1. "The State of the Environment in OECD Member Countries," Paris, 1979.

sectors, such as energy, there had been an improvement in the relationship between resource consumption and output and also an improvement in the relationship between emissions of pollutants and output. Ministers concluded that in the long run environment protection and economic development are not only compatible but interdependent and mutually reinforcing. Ministers also requested that further studies on the social and economic benefits of environment policy be pursued in the framework of the OECD.

4. Ministers, on the other hand, noted that environmental quality remains a source of concern in certain fields, such as the supply and quality of water including, in particular areas, drinking water; the adverse effect of the excessive use of fertilizers; the impact on ecosystems of the long range transport of airborne pollutants; air pollution by recently recognised pollutants; increasing noise levels in some areas; the risk of accidents involving hazardous substances; and, in general, the growing evidence of linkage between environmental deterioration and human health. The Ministers expressed their determination to take the necessary measures in their national environmental policies and agreed to actively pursue their co-operation within the OECD in these areas.

5. Focussing on environmental policies for the 1980s, Ministers agreed that, even if further economic developments were only moderate, the potential environmental consequences in many sectors could be considerable. For this reason there should be no relaxation of effort. Indeed strengthened measures, designed with a view to their economic efficiency and cost effectiveness and focussed on prevention rather than remedial action, will be needed to maintain and improve present levels of environmental quality; and to promote the concept of qualitative growth - a partnership between environmental and economic concerns. Moreover environmental considerations must be integrated at an early stage in all major decisions likely to have significant impacts on the environment. One means of achieving this is through procedures for the assessments of projects along the lines covered by a new Recommendation which has been approved by Ministers.

6. Recognising the need for vigorous, forward-looking environmental policies, Ministers adopted unanimously in the name of their governments a Declaration[1] setting out the key elements of policies which are needed in the coming decade.

7. Ministers discussed a number of specific issues which require action at the international level.

i) Recognising the need to reach a better balance of energy supply and demand, they stressed that greater weight should be given to environmentally beneficial options such as conservation and the development of new forms of energy sources. However,

1. See page 9.

such sources are several decades away and Ministers called for vigorous measures to control pollution in the transition period. They recognised also that among other sources available coal will be an important source of energy in the short and medium term, and they approved a Recommendation to deal with the environmental implication of increased production and use of coal.

ii) With regard to the development of nuclear power which will become an important source of energy in some countries Ministers stressed the need to collect and provide to the public as complete information as possible on safety requirements and environmental implications.

iii) They welcomed steps within Member countries and the Organisation to investigate fully and inform the public of the implications of the 1979 accident in Harrisburg and to assess the environmental implications of all energy sources, including nuclear.

iv) Ministers recognised the need to reinforce measures to protect the coastal and marine environment against oil prollution. They stressed the need for preventive action and the benefits to be derived from international co-operation in this area.

v) Ministers noted that international tourist movements have increased tenfold over the last 30 years and that this rapid increase is expected to continue at a rate of 6 per cent per annum over the next five years. Since this growth takes place on an essentially fixed base of assets, in terms of natural and historic resources, and puts stress on the environment in many areas, Ministers agreed on a Recommendation on policies to be followed to reconcile tourism and environmental needs.

vi) Many forms of environmental pollution can, in one way or another, be traced to chemicals. Ministers recognised the all-pervasive importance of chemicals in everyday life and the economic significance of the chemical industry; they stressed at the same time that anticipatory policies were essential to keep chemical pollution under control and they renewed their support for the OECD programmes on control of chemical substances to protect the environment and human health and to harmonise policies while avoiding undue negative effects for the economy and trade distortions. Many Ministers called for an early meeting at high level to review the implementation of the programmes.

vii) Ministers referred to the important international problems arising between Member countries because of transfrontier pollution, more particularly in connection with long-range transboundary transport of airborne pollutants and the pollution of international rivers. They pointed to the progress made since 1974 in developing international co-operation in this area, but recognised that further international efforts were still required.

viii) In view of the importance for the environment of urban planning and management, Ministers expressed their satisfaction concerning the work of the Organisation in the past and welcomed the proposed new programme in this field.

8. Ministers stressed the need to improve the information basis on the state of the environment, on the impact of relevant activities and on environmental policies themselves. In order to perform this task in an increasingly effective manner they agreed on a Recommendation calling for systematic monitoring and reporting on the evolution of the state of the environment in Member countries.

9. Ministers expressed their determination to strengthen their co-operation in the OECD framework with regard to environmental issues which are both economically and internationally significant. Mindful also of the responsibility shared by all countries to protect and enhance the environment on a worldwide basis, they stressed the importance they attach to co-operation with countries with different economic systems or at various stages of development.

DECLARATION ON ANTICIPATORY ENVIRONMENTAL POLICIES

The Governments of OECD Member countries[1] and of Yugoslavia.

a) Recalling their Declaration on Environmental Policy adopted in 1974;

b) Noting the significant achievements over the past decade in reducing pollution, conserving natural resources and providing a better human environment;

c) Recognising the need to integrate environmental, economic and social policies;

d) Convinced that improving the human environment involves sustained long-term effort requiring policies that take into account at an early stage the environmental consequences of major decisions;

e) Mindful of the hazards to health and the environment arising from certain socio-economic activities;

f) Aware that market mechanisms by themselves often do not induce decisions that reflect their environmental consequences and costs;

g) Determined to pursue vigorous environmental policies in a manner consistent with their constitutional, legal and market economy systems;

DECLARE THAT:

1. They will strive to ensure that environmental considerations are incorporated at an early stage of any decision in all economic and social sectors likely to have significant environmental consequences.

2. They will seek more effective institutional, economic and other means to integrate environmental policy with policies in other sectors and in so doing, will attach priority to land-use planning and to the chemicals, energy and other sectors having a major impact on the environment.

---

1. The mention of "Governments" is deemed to apply also to the European Communities.

3. They will, where appropriate and possible, employ economic and fiscal instruments, in combination as need be with regulatory instruments, to induce public and private enterprises and individuals to anticipate the environmental consequences of their actions and take them into account in their decisions.

4. They will encourage the design, development and use of processes, products and urban form that conserve resources and energy, and that protect and enhance the environment.

5. They will, in proposing laws and making regulations, seek to avoid unduly complex or conflicting requirements and unnecessary delays in decisions affecting the environment.

6. They will endeavour, to the extent practicable, to develop systems to account for changes in environmental quality and related resource stocks.

7. They will encourage public participation, where possible in the preparation of decisions with significant environmental consequences, inter alia, by providing as appropriate information on the risks, costs and benefits associated with the decisions.

8. They will support the promotion of environmental objectives and awareness in the field of education.

9. They will strengthen their co-operation in the OECD in order to encourage arrangements between interested parties on international environmental problems, and to seek harmonization of national environmental policies.

10. They will continue to co-operate to the greatest extent possible, both bilaterally and through appropriate international organisations, with all countries, in particular developing countries in order to assist in preventing environmental deterioration.

II

ANTICIPATORY ENVIRONMENTAL POLICIES

(A Discussion Paper)

Background

1. OECD Environmental Ministers met in Paris on May 7th and 8th, 1979, against a background that is at once similar and markedly different from that which existed at the time of their previous meeting in 1974. Several features stand out.

2. Public support for measures to improve the environment remains high. According to recent polls in several countries, it stands as one of the three highest priority concerns of the public along with inflation and unemployment, and it continues to grow rapidly.

3. The field of environmental action is expanding. Pollution abatement remains a priority but measures to conserve resources and improve the quality of life command increasing attention. Policies that are anticipatory and preventive are receiving increased emphasis, reinforcing policies that are essentially reactive and curative. Fundamental questions about alternative growth patterns and lifestyles are being raised.

4. The integration of environment concerns in major investment decisions has become urgent, especially for agriculture, forestry, energy, chemicals, tourism, transport and urban development. There is a growing time-lag in decisions and a growing number of deadlock situations, often concerning the siting of large-scale, high impact developments having long-term life cycles. In energy alone, some studies suggest the need for a start on over 250 new sites for power stations to meet the needs of the OECD region through 1985.

5. Pressure for greater participation in decision-making remains strong and will probably increase, giving rise to a wide range of complex issues which will impact upon development, environment and society as a whole.

Changing Economic Conditions

6. As Environment Ministers of OECD Member countries met again after four years, one factor stood out above the rest : the change in the

economic situation. Economic trends to 1985 give rise to concern, and provide a new and difficult set of conditions for environment policy.

7. Environment policies are increasingly perceived to be justified on economic, as well as social and ecological grounds. National studies, based now on several years of experience, suggest that total benefits are well in excess of the total costs of measures to abate pollution. Indeed, the present situation, wherein expenditures on pollution amount to 1-2 per cent of GNP and damage from pollution ranges between 3-5 per cent of GNP, entails a welfare loss and justifies an increase in expenditures on environment. Recent polls in some countries indicate that increases for this purpose would command broad public support.

8. The environmental consequences of economic growth in the coming years will be far from negligible, even though that growth will continue to be slower than in the past. Total pollution emissions in a number of sectors, assuming present quality standards and technology, could rise 30 per cent or more between 1978 and 1985. If present environmental standards are not achieved, a significant deterioration of environment quality could take place at local, regional and international level. If environmental measures were to be relaxed, deterioration would be greater. In the face of growth, even slow growth, only strengthened measures will maintain and gradually improve environmental quality.

9. These will be difficult to achieve. Slower growth, with a slower replacement of capital, can reduce the rate at which newer, less-polluting processes find their way into the system. Marginal firms, or firms facing intense competition, may find it difficult to transfer investment to pollution control measures. Major structural and sectoral shifts, and shifts in the pattern of international economic relations, can affect entire regions and their capacity to sustain environmental programmes. Slower growth in public expenditures can increase the competition for government support of environment measures.

10. Equally significant, slower growth with continued higher-than-acceptable rates of inflation creates a climate of credibility for claims that the cost impact of environmental measures is severe. Studies since 1970 demonstrate clearly, however, that this is not the case: the macro impact of environment measures on inflation, employment, productivity, investment and balance of payments, is negligible and not always negative. Yet, it can be expected that proposals for new environmental measures will be subject to increased scrutiny against not only their potential economic and social impact, but also their efficiency and cost-effectiveness.

Anticipatory Environment Policies

11. This new situation calls for sensitive and determined action on the part of Environment Ministers. It poses a challenge to sustain the achievements gained over the past decade, and to develop a new generation of policies, aimed both at anticipating and preventing environmental deterioration from occurring and at creating conditions in which the growing demand for environmental goods can be met effectively and efficiently.

Scope

12. Anticipatory policies would embrace a broader field of environmental action: not only pollution control, but also the conservation of resources, including our historical and cultural patrimony, and the provision of amenities and other measures to improve the quality of life. It could even extend into the realm of alternative patterns of growth, consumption and changing life styles.

Nature

13. Formulated within a longer-term perspective, anticipatory policies would include at least two types of environmental action: those which plan positively to create the environmental benefits that society wants, and is increasingly prepared to pay for; and those which seek to prevent the environmental degradation that, in their absence, can predictably result from human activities.

13.1. In considering actions of the first type, Ministers will be concerned with policies that actively promote human health and well-being; that serve to protect other environment species and natural ecosystems; that influence resource-conserving urban patterns, energy and transportation facilities; and that create amenities, conserve landscapes and the built heritage.

13.1. In considering actions of the second type, Ministers will be concerned with policies that reduce the production, marketing and disposal of materials and products dangerous to the environment; that encourage integrated, closed-cycle industrial processes which minimise waste; that encourage recycling and the design of products that facilitate their ultimate recycling; and that make economic use of waste residuals.

14. Generally, anticipatory policies involve actions to ensure that environmental requirements are taken fully into account at the earliest possible stage of any major decision affecting the environment. Considering the complete product cycle from design and development

through to marketing, use and disposal they would tend to focus on the forward design and planning rather than on the use and waste disposal end. This would be true whether the cycle concerns energy, a new commercial product or a new urban development or re-development. This is the focus, for example, of the new OECD Special Programme on the Control of Chemicals launched in September, 1978.

15. Anticipatory policies are intended to reinforce reactive-curative policies, not to replace them. Indeed, where changes in environmental quality can be achieved by measures that are simple to apply, involve small-scale technology, and are likely to be effective over a short time-frame, there seems little need to be concerned about anticipatory actions: market mechanisms or quick-acting regulation can be relied upon. Similarly, reactive measures, will continue to be necessary to cope with environmental accidents.

Instruments

16. Environmental action of an anticipatory nature can be pursued through a wide range of instruments. They include: technology assessment, environmental impact assessments, land use planning laws, design and product regulation and controls of various kinds. But instruments that are essentially economic in character can also be relied upon and may, in some cases, prove more acceptable to the public in that, properly used, they leave more decisions open to individual choice than do regulations. Such instruments include taxes, charges and financial incentives used to encourage choices compatible with the maintenance of good environmental conditions. In this regard, regular assessment of the environmental implications of proposed changes in tax and fiscal policies seems critical.

Advantages

17. The economic and social arguments favouring anticipatory actions are often convincing. With regard to preventive actions, the cost-effectiveness of abatement measures built into the design of products and processes is much higher than that of measures requiring post-hoc prescription, i.e. re-design and retrofitting. Often, they are quite profitable to the industry concerned. In addition, society avoids the heavy external economic, social and health costs associated with the damage effects of the pollution in question. Similarly, with regard to positive planning measures, society can avoid the high costs associated with environmental mistakes, costs that can represent a heavy annual, unproductive burden on the nation. Energy and resource consumptive urban forms, for example, can impose a continuing drain on a nation's balance of payments. Once developed, however, urban

forms can be changed only slowly. The economic and social arguments, of course, are often supported by critical ecological factors, stemming from phenomena such as irreversibility, bio-accumulation, persistence and long-term health and genetic effects.

Participation

18. Anticipatory actions may carry further advantages in the area of public participation, providing the time necessary for the appropriate authorities to interact with the public concerned in a pending decision and to take their views into account in defining and selecting options.

19. There is little doubt that to date public participation has led to significant and beneficial results in many areas. It has often been a source of strong political support for the adoption of environment measures. Equally, there is little doubt that such participation has led on occasions to considerable delays in decision-taking and consequent increased costs. These costs need to be compared, however, not with the decision-costs of previous eras without adequate environmental protection laws, but rather with the higher costs that result when the views of the public directly affected by a major decision have not been sought.

20. The demands for participation can be expected to increase with higher levels of education, income, leisure time and information. It raises difficult questions about the public's proper role in decision-making and about policies and mechanisms to make that role as effective and productive as possible.

Research and Development

21. Anticipatory policies depend on augmenting the stock of scientific technical, economic and socio-political information on environmental issues. It also requires the development of effective clean technologies to reduce pollution, to prevent the wastage of scarce resources and to improve the quality of life. It is therefore of paramount importance that environmental research and development be strengthened.

Difficulties

22. Anticipatory actions, nevertheless, are often difficult and sometimes impossible to adopt. This is essentially because they are long-term, because they are, by definition, introduced before damage to health and environment has created a demand for amelioration, and because they often involve front-end-investments. This reflects political reality. It also reflects the fact that while, during the past decade

of environmental action, society has learned to deal, at least in part, with economic externalities, it remains incapable of dealing with time-externalities (i. e. adverse effects falling on future subjects, or discounting the future).

Integration

23. It is evident that, for political and other reasons, anticipatory actions require the closest integration of environment policy with economic and social policy and policies in other sectors; such as principally energy, health, industry, land-use, tourism and transportation. In other words, actions in support of anticipatory policies cannot be left entirely to environment Ministers but must also involve other Ministers; often the required action can be taken only by them.

Institutional Arrangements

24. Thus, anticipatory environment policies may require special institutional and other arrangements to ensure continuing interaction, with other Ministries and agencies, public and private, especially at the level of fundamental options. It would seem important, for example, that means exist to ensure that:

- Environment Ministers are automatically informed of policies and plans in other fields that have, or may have, significant environmental impacts;
- Suitable mechanisms are developed to ensure that environmental considerations are brought to the centre of major decision-taking processes;
- Better instruments are found to encourage public participation and make it more informed and effective;
- Environment and other Ministers are informed of evolving public views concerning the environment.

International Co-operation

25. Anticipatory actions, especially those of a preventive nature focussed on the forward design and planning end of the design-production-marketing-use cycle, would appear to call for a strengthening of international co-operation. The OECD Special Programme on the Control of Chemicals is the latest case-in-point. Without such international co-operation, governments may be reluctant to develop and apply environmental design and product standards for fear of affecting the international competitive position of their national industry. Smaller countries may be unable even to consider it. Without it, too, the introduction of national design and product standards could lead to trade distortions and non-tariff barriers.

26. Anticipatory actions depend in large measure on the ability to forsee at least roughly the environmental implications of likely changes in economic and social conditions, including the terms of trade, as well as emerging developments in the area of environmental science and technology. International exchange of information, analysis and experience to these ends is of growing importance.

27. Increased emphasis on anticipatory actions, therefore, will call for gradual strengthening of mechanisms to notify, consult on, and review actions proposed by Member countries in the environmental field. Furthermore, the on-going environment programme of the OECD developing policy options and guidelines, should be focused more on longer-term issues than lend themselves to anticipatory actions as well as on emerging issues and developments in the field of environmental science and technology.

III

ENVIRONMENTAL POLICIES AND ECONOMIC PROSPECTS FOR THE 1980s

SUMMARY

The aim of this Chapter is to examine, in the light of the likely economic prospects up to the mid 1980s the need for, and possible obstacles to, environmental policies, in particular pollution prevention and control measures. For this purpose, it first describes a broad economic scenario for the coming 5 years; it then examines the likely impact of this scenario on the environment; third, it analyses the question of compatibility between environmental and economic objectives; and, finally, it reviews the prospects and possibilities for an improvement and strengthening of environmental policies.

The economic scenario up to the mid 1980s is strongly influenced by economic events since 1974. In contrast to the rapid economic growth of the 1960s and early 1970s - an OECD average of 5 1/2 per cent per annum, full employment and price stability - the outlook is less promising. An ambitious but feasible target for the coming period would be an average economic growth rate of around 4 1/2 per cent for the OECD as a whole, but this would require strong policy initiatives at home and high degree of international co-ordination. Even so unemployment levels and inflation rates would remain, on the average, higher than in the 1960s.

The environmental consequences of projected economic development in the coming years will not be negligible. In the absence of improved pollution control performance, associated in most cases with tighter environmental policies, deterioration of environmental quality could occur over the period to 1985.

i) The overall picture associated with the industrial sector poses special challenges. In the absence of tighter policies for pollution control and the development of appropriate strategies for environmentally favourable industrial growth, there could be under certain assumptions a 10 - 25% increase in pollution emissions for the major conventional pollutants between 1978 and 1985 and a greater increase for the less-easily controlled micro-pollutants.

ii) For the same period, certain types of pollution associated with agriculture could, in the absence of changed patterns of demand and use, increase by as much as 50%. This issue is all the more serious since the non-point source of this pollution makes it difficult to control.

iii) Similarly major policy efforts are needed to prevent environmental deterioration from arising in two other major polluting sectors: energy and transport. Indications point toward degradation of environmental quality in urban and the tourist areas unless policies are undertaken no to pre-empt such developments.

The economic implications of environmental policies are the costs that have to be paid for the benefits these policies produce. Available evidence suggests that so far these costs have been small in terms of economic aggregates.

i) The global employment effects of environmental programmes in the short-run appear so far to have brought a net creation of jobs. This net positive total, however, masks gains and losses from one region or one industry to another. The longer run effects are less well known.

ii) The inflationary impact of environmental policies on the average has been around 0.2 - 0.3 per cent annually as measured by the consumer price index. The implementation of the PPP[1] has had an overall favourable effect on the price structure by inducing increased efficiency.

iii) Governmental environmental expenditures are a small part of total public expenditure, averaging less than 1 per cent. Nevertheless, in a period of increased competition between various claims for public funds, expenditures on environmental goods will have to be justified in terms of their cost-effectiveness and their benefits. It might be also necessary and desirable to cover a larger proportion of such expenditures through user charges and fees.

iv) Given the need for investment funds, and for higher profitability in the private sector, OECD economies are striving to improve the rates of productivity growth which have been in decline for some period. The positive contribution of environmental policies to productivity, in form of better health, reduced manhour losses and technological improvement, have been so far under-estimated. Partly as a result of this it has been asserted that environmental policies have to date produced a negative effect on productivity.

v) In the field of foreign trade and international capital flows international co-operation has tended to minimise the potential negative effects of environmental policies. The present on-going work on chemicals is a good example of international efforts in this field.

1. Polluter Pays Principle.

In summary, the evidence suggests that the economic impact of environmental measures has not to date presented a conflict between environmental and economic objectives and the policies needed to achieve them.

Evidence from a number of OECD countries indicates that public support for environmental measures remains strong despite slower economic growth and unemployment. This stems from a number of factors, amongst them higher incomes, a greater awareness of environmental needs and growing recognition of the benefits of environmental measures.

It is also suggested that in the future environmental measures will be subjected to greater scrutinu on the part of those responsible for balancing competing claims for scarce resources. They will therefore be a greater need to evaluate the future benefits of proposed measures. To date many of the more important benefits have proved difficult to measure: for example, improvements in health and savings in health expenditures, a more pleasant built and natural environment improved labour productivity, improved productivity of land and significant improvements in technology and innovation in production processes.

The course of future environmental policies, will be guided by needs as perceived by the governments, by public preference for environmental goods and by the availability of resources. Resources for environmental measures are not likely to be as available as in the past and proposals are made on a number of ways to strengthen and improve the efficiency and effectiveness of environmental policies. In addition, serious consideration should be given to closer integration of environmental policies with policies in other sectors.

Future environmental policies, should seek to ensure that environmental requirements are taken into account at the earliest possible stage of major decisions affecting the environment. This could ne achieved partly through greater use of economic instruments side by side with more traditional regulatory measures.

i) In order to facilitate structural changes towards less polluting industries and environmentally favourable products the consistent application of the Polluter-Pays Principle is advocated. The inclusion of the cost of environmental protection in the cost of production should serve to accelerate desirable structural changes as well as to reduce the need for subsidies.

ii) The financing of environmental programmes in the public sector would be facilitated through greater use of emission charges and user fees. Such a policy would not only provide the funds to cover associated costs, but could also induce less polluting behaviours.

iii) New regulatory measures will continue to be necessary but great care should be taken to avoid unnecessary delays associated with the regulatory processes.

iv) Regulatory and economic instruments both have often served to induce such innovations as have research and development programmes aimed specifically at developing abatement technologies. Such programmes should be intensified in the future.

v) The integration of environmental policy with economic and social policy and policies in other sectors, such as energy, land-use, transportation, etc. will be more essential in the future. Various areas where such integration would be most fruitful are set out.

In conclusion the economic outlook for the period up to 1985 foreshadows the need for vigorous environmental policies. This will be true even assuming a relatively low economic growth scenario. On the other hand, the economic pressures on environmental policies could be significantly greater than in the past. Given this combination of circumstances environmental measures of an anticipatory and preventive nature will be required not only to meet future environmental demands but also to ensure greater economic efficiency.

INTRODUCTION

1. The Issue

Following the economic events of 1974-75 and the subsequent economic stagnation, economic growth has come into focus again. At the national level unemployment, inflation, balance of payments and investment have been in the forefront of discussion. Governments and private industry have also shown considerable concern with productivity and profitability. The discussions have raised the question of reconciliation between economic growth objectives and other social, industrial and environmental aims. While the need for the continuation and even strengthening of environmental measures has been clearly recognised, measures have been taken in some OECD countries to ensure that the implementation of environmental policies is adjusted to harmonise more closely, than in the past, with economic policy objectives. In other cases environmental goals proposed for the early 1980s are now being re-examined in the light of the changed economic climate. In view of these rapidly and radically changing circumstances it is proposed to examine if and to what extent economic and environmental goals conflict and to what extent and how they can be reconciled during the period up to 1985.

Because of the broad horizon and complexity of the problem the paper covers, it has been necessary to introduce a number of simplifications. Consequently, the discussion is subject to certain qualifications:

i) The OECD countries are dealt with in the paper by necessity as a single unit and the discussion in most of the paper is in terms of averages or based on information from countries with relatively well developed and well documented environmental programmes. This is a simplification, because there are clearly significant differences not only between countries, but also regions and population groups.

ii) The environmental concern is meant relatively broadly but in fact the information available and therefore the discussion is largely limited the pollution abatement. Issues such as scarcity of resources, conservation and working environment, which are likely to be important issues in the future, are not dealt with explicitly.

iii) Because of the difficulty in generalising micro-information at the level of the firm, industries or regions the paper deals with both the economic and pollution problems in macro terms.

In spite of these reservations there is a broad consensus that a discussion in the context presented in this paper will be an important contribution to the formulation and implementation of environmental policies for the earlier part of the coming decade.

2. The Economic Outlook[1]

This section briefly examines the overall economic climate which is likely to prevail between the present and the mid-1980s. Economic developments in the medium term will be of crucial importance to environmental concerns. On the one hand, they will have direct consequences on the environment by, inter alia, creating new pressures and problems. On the other, they will influence the overall social and economic climate against which environmental targets and policies will have to be formulated.

The present economic situation is well known. Most recent OECD estimates and short-term forecasts suggest that our economies have still not fully recovered from the 1974-75 crisis. The growth in output in 1979 is not expected to be above 3 per cent in the area as a whole,[2]

1. The scenario in this paper is largely based on "Economic Outlook" OECD, December, 1978, and an unpublished OECD document.

2. GROWTH OF REAL GNP IN THE OECD AREA

Percentage changes seasonally adjusted at annual rates

	AVERAGE 1959-60 to 1972-73	1965-66 to 1975-76	FROM PREVIOUS YEAR		
			1977	1978	1979
United States	4.2	2.7	4.9	3 3/4	2
Japan	10.9	8.2	5.2	5 3/4	4 3/4
Germany	4.9	3.3	2.6	3	4
France	5.9	4.7	1.6	3	2 1/4
United Kingdom	3.3	2.1	2.7	3 1/2	4
Canada	5.1	4.7	1.7	2	3 1/2
Italy	5.6	4.1	4.0	3 3/4	3
Total of above countries ..	5.5	4.0	1.8	2 1/4	3
Other OECD countries	5.5	4.4	3.7	3 1/2	3
Total OECD	5.5	4.1			

Source:"OECD, Economic Outlook," December, 1978 and July 1976.

well below the 5 per cent rates achieved in the 1960s and early 1970s and possibly insufficient to keep unemployment from rising. Among the most disturbing recent trends has been a slowdown in growth rates of labour productivity. Resumption of faster productivity growth could influence the prospects for most governmental including environmental programmes in many Member countries. And while inflation is decelerating its rate (7.9 per cent for 1978) is still above desirable levels with wide inter-country differences being particularly disquieting.

Finally, though the pattern of world balance of payments is moving towards a better equilibrium, this external constraint remains a very important one in many countries and, together with inflation, limits the scope that individual governments have for new policy actions.

The economic outlook for the coming half decade is strongly coloured by the present situation. A continuation of present trends cannot be ruled out. On the other hand possibilities exist for an imaginative and internationally coordinated policy response which could lead to a recovery in the world economy and bring the OECD area back to full employment.

A continuation of the present cautious stance of demand management policies, induced by fears of renewed inflationary pressure and/or balance of payments difficulties, could result in growth rates of overall output in the medium-term close to those recorded in the recent year. Yet it is not clear that such a low scenario would necessarily overcome some, or even most, of the constraints which today face policy makers. For one thing, a depressed economic climate would, by lowering investment levels, lead to a longer run deceleration of the potential growth of our economies. For another, inflation need not decelerate despite lower demand pressures if the conflict over income shares were aggravated, as it might well be, in a situation of only slowly rising real wages. Finally, international payments imbalances would almost certainly persist while unemployment would be even above today's high levels.

An alternative, and more desirable, outcome would be an acceleration of growth from the present OECD average of 3 per cent to a feasible rate of 4 1/2 per cent per annum. This would permit a gradual absorption of slack and a return to full employment by the mid 1980's. While the inflation rate thought to be consistent with this scenario would remain relatively high by historical standards (6.5 per cent between now and 1985), it would be declining through the period to a more acceptable 5 per cent in the terminal year.

While this scenario is feasible, it is also ambitious . It would require a large degree of international coordination, to alleviate potential balance of payments problems, it would necessitate strong policy initiatives at home to overcome domestic constraints and it would inevitably involve considerable risks. In particular:

i) Domestic policies would have to ensure a revival in private investment. This would require not only a progressive increase in demand and a shift from private and public consumption towards private capital formation; but also in addition a more than cyclical shift in income distribution towards profits;

ii) At the same time, policies would have to be suitably differentiated internationally, particularly in the earlier stages of the upswing so that the external constraints facing weaker countries be relieved;

iii) On the risks side, an acceleration in demand growth could have unfavourable consequences on the inflation rate, particularly in a situation in which inflationary expectations are still rife and commodity prices particularly volatile;

iv) Nor would realisation of the scenario solve all problems simultaneously; unemployment levels even if gradually reduced, would remain high over the period as a whole.

The major conclusion which can be drawn from this bird's eye view of possible medium-term developments is that the economic situation is likely to remain unsatisfactory. Even under favourable assumptions, growth rates of output would not return to those witnessed in the 1960s, while unemployment levels and inflation rates would remain, on average, higher than in that period. And if economic policies were unable to meet fully the challenges put to them, the likely outcome would be less favourable, with the industrialised world continuing to limp along a path of relatively low growth and unsatisfactory overall performance.

Additionally, there is now an awakened recognition that the events and growth patterns of the fifties and sixties were characterised by specific factors:

- low and sometimes declining raw material costs
- expanding populations
- rapid increases in productivity in agriculture and manufacturing industries made possible by increased urbanisation, education and major technological innovations.

Towards the end of the seventies and probably for the remainder of this century our economies are facing conditions that would be less supportive of growth, in the traditional sense of the word, in the OECD countries. The constraints include:

- rising costs of raw materials
- stabilising populations
- smaller productivity increases due to: reduced possibilities for scale economies in industrial and agricultural production, a slowing down in major technological innovations, and structural

change, such as the increasing share of the service sector in the OECD economies.

And above all there are signs in our societies of a desire for a better balance between quantitative and qualitative growth objectives.

1

IMPLICATIONS FOR ENVIRONMENTAL QUALITY

Economic growth inevitably exerts pressure on environmental quality. Materials extraction and the production and consumption of goods and services result in discharges of many polluting substances to the environment. The relationship between economic growth and environmental quality is complex and depends critically upon the nature and stringency of environmental protection programmes adopted by governments and the development of pollution control technology. It is nevertheless useful to examine closer, and at a broad international level, the environmental implications of the course of economic outlook discussed above.

In the following sections scenarios are offered of likely environmental pressures in the period to 1985 based on the assumption that present discharge standards will continue. These are not offered as forecasts - clearly the levels achieved will depend upon the formulation and strengthening of environmental programmes adopted by Member governments - but they should provide a baseline against which the need for more stringent controls can be evaluated and the present challenges can be seen.

Economic growth over the period to 1985, even slow economic growth, will mean extra pressures on the environment. Assuming a growth in emissions proportional to growth in GNP, a 3 per cent growth rate from 1978 to 1985 would imply a 23 per cent increase in emission and a 4 per cent growth rate would imply a 36 per cent increase.[1]

Fortunately, certain factors act to offset this proportional growth in total emissions:

i) The replacement of old capital equipment and the introduction of new equipment incorporating technological improvements constructed to meet current emission standards mean that emissions increase less than proportionately with output.

---

1. Note should be taken that the relationships between emissions and both ambient concentrations and environmental damage are complex, and not necessarily proportional. However, because of lack of data changes in total emissions are often used as useful indicators of changes in environmental quality of damage.

The impact of such potential improvements will depend heavily on the rate of replacement of capital and new investment. Technological progress combined with a high rate of investment could significantly reduce and even offset to a significant extent the growth in environmental degradation.

ii) The adoption by Member countries of increasingly more stringent pollution control policies would also favourably affect emission-production ratios. Additionally, environmental policies affecting the location of polluting activities can have a mitigating influence on environmental degradation.

iii) Changes in the structure of production and consumption also impact upon the environment. These may or may not all be favourable in nature, e.g. shifts in industrial structure towards or away from less polluting industries, shifts in consumer taste for less environmentally damaging goods.

In order to find out how these impacts combine in the mid-term future, a closer sector-by-sector analysis is required.

1.1. Industry

The examination of recent trends in the industrial structure, helps to formulate some general conclusions to be made about trends in industrial growth in OECD Member countries over the period to 1985. The trends point to a growth in industrial production continuing at a slightly higher rate than growth of GNP. The existing forecasts, however, indicate certain structural changes occuring within the industry sector. These changes stem from the differing growth rates in the various sub-sectors caused in turn by changes in the pattern of final demand and/or by changes dictated by international markets.[1]

A country-by-country review of the anticipated structural changes shows that three sectors are expected to increase their shares of to industrial production: the chemical, electrical engineering, and mechanical engineering industries. Other industries, such as food processing, textile and wearing apparel (including leather and footwear), and in a number of cases, steel and ship building, are expected to decrease their shares of industrial production. In practically all cases, the decline will be only relative, not absolute.

---

1. Environment policies can also affect industrial structures; e.g. the application of the Polluter Pays Principle (Section 2.2) expected to favour less polluting industries. In the longer term, beyond 1985, more fundamental changes in industrial and social structures may be expected; the consideration of trends in the longer run toward low-pollution and low-energy use, so called 'post-industrial', society is beyond the scope and time-frame of this paper.

What will be the impact of such structural changes on pollution?

As far as the important conventional pollutants, are concerned (such as oxidisable substances, suspended solids, particulates and SO_2) the impact is not likely to be very significant; the dominant factor will be industrial growth.[1] The autonomous changes in industrial structure alone are unlikely to have a significant offsetting influence on increases in pollution. Pollution-intensive industries are to be found in declining as well as in rising industries.

However, the impact of structural changes on micro-pollutants (such as heavy metals and synthetic chemical compounds) could be more important. The environmental threat associated with the chemical residuals of industrial processes have become widely recognised in recent years and appropriate pollution control technology has been developed. Nevertheless, micro-pollutants are a less tractable form of industrial pollution than conventional pollutants. In particular micro-pollutants are associated with the chemical industry and segments of the electrical engineering and mechanical engineering industries; these are sectors expected to increase their shares.

It is difficult to estimate the impact of capital replacement. Clearly, some old, obsolete, and polluting production units will be closed down and replaced by modern and less polluting units. Prescribed emission standards can usually be satisfied with new equipment at lower costs than with older equipment. Insofar as there is capital replacement, and the embodied technological developments are less polluting, there will be an off setting impact upon total emission levels.[2]

The overall picture associated with the industrial sector suggests continuing challenges. In the absence of new and reinforced policies for pollution control, and the development of appropriate strategies for environmentally favourable industrial growth, there could be under certain assumptions a 10 - 25 per cent increase in pollution emissions for the major conventional pollutants between 1978 and 1985 and a greater increase for the less controlled micro-pollutants.

---

1. Using available pollution coefficients and data on industrial structure relating to Japan for the period 1970-1985, OECD analysis indicated that pollution levels, as measured by SO_x, BOD, and waste were affected only slightly by a change in structure. The size of industrial growth dominated the growth in total emissions.

2. The contribution this will make to offsetting the potential growth in emission associated with the growth of the industrial sector in the period to 1985 (e.g. 23-26 per cent) is difficult to assess in the absence of accurate information on the pollution control effectiveness of new equipment. This in any case will be to a large extent determined by the nature and stringency of pollution control policies adopted by countries over the corresponding period.

On the assumption, however, of 40 per cent capital replacement over that period and that this new equipment will be 10-20 per cent less polluting than increases in total emissions associated with industrial growth to 1985 would be 10-25 per cent.

1.2. Agriculture

The picture associated with projected trends in agriculture also presents problems requiring attention. It is thought that there will be no physical or technical obstacles to the expansion of agricultural production, which is expected to grow at about the same rates as in the past.[1] This assumes that relative prices for agricultural products will be maintained and that technical innovation will continue.

In the past, agricultural production has been increasing (in volume, which is relevant for the environment) at a rate of 2-3 per cent per year. This increase which was achieved with decreasing labour and constant land is accounted for by increased capital, by technical progress (new types of high yield plants were introduced) and above all by increased use of artificial fertilizers: over the 1965-75 period, the use of nitrogenous fertilizers increased about 7 per cent per year.[2]

Continuation of those trends implies environmental degradation. Agricultural and forestry activities do not contribute much to air pollution, except in local circumstances. But they can be major contributors to water pollution; in the U.S.A., for instance, it has been estimated that these activities are responsible for 40 per cent of BOD, of suspended solids and 33 per cent of dissolved solids in water.[3] Forestry activities are important in a number of Member countries and increasing and more intensive timber production will require careful balancing of economic and conservation objectives. In the case of agriculture, run-off from agricultural land of phosphate and nitrogen fertilizers is a major cause of the eutrophication of inland waterways. The case is probably worse with specific pollutants. Trace organic chemical pesticides are found in many waterways, and may pose a serious threat even at low concentrations because of the bioaccumulative nature of some of these compounds. This raises serious health as well as economic problems for public water supplies. As many of these trace chemical compounds are not readily decomposable biologically the increasing residual stock in agricultural land and waterways is an issue of concern. The development, however, of synthetic chemical pesticides which are less persistent and less environmentally damaging will act to offset these threats.

Major pollution problems associated with agriculture are linked mainly to fertilizers and pesticides and these are projected

---

1. Study in Trends in World Supply and Demand of Major Agricultural Commodities, OECD, 1976.

2. State of the Environment Report, OECD, 1979.

3. "Economics and the Environment", A.V. Kneese, Penguin, London, U.K., 1977, p. 48.

to increase at rates of growth higher than that of agricultural output; an average 4-6 per cent annual growth is projected in consumption over the period to 1985.[1]

Consequently between 1978 and 1985, certain types of pollution associated with agriculture could, in the absence of changed patterns of demand and use, increase by as much as 50 per cent. The issue is all the more serious since the non-point source of this pollution makes it difficult to control.

1.3. Transport

In transportation, the major role is played by automotive vehicles. The annual increase in the stock of vehicles, which was very rapid in the past (4 per cent in North America, 8 per cent in OECD Europe, and 22 per cent in Japan, over the 1965-75 period) will slow down, but is expected to remain significant (- 1.7 per cent in the U.S.A., less than 4 per cent in EEC countries, 5 per cent in Japan and 10 per cent in some other OECD countries such as Portugal, Spain, Greece, Yugoslavia, Turkey) for the 1975-1985 period.[2]

Mobility, defined as the average number of kilometers travelled per car, is also expected to increase due to i) increases in household expenditures for transport in Europe and Japan ii) the continuing sprawl of settlements inducing larger commuting distances and iii) trends in consumption patterns resulting in more time for and emphasis on tourism and leisure activities. A rate of increase of nearly 7 per cent per year over the 1975-85 period has been forecasted for EEC countries.[3] Much lower rates are likely for North America and Japan, but higher rates can be predicted for the less industrialised countries of the OECD.

The consequences for emissions depends upon the relative magnitude of three different parameters: the increase in the number of vehicles, the increase in mobility and the rate of replacement of old vehicles by new and less polluting ones (automobiles produced today are much less polluting than automobiles produced previously and than the average automobiles in use today). In countries where mobility and the number of vehicles will grow slowly (like the United States) the replacement factor will predominate so that emissions could stabilize, or even decrease if emission standards are strengthened. In other countries experiencing

---

1. See USDA Report," Evaluation of pesticide Supply and Demand 1977; Wood-Mackenzie Pesticide Industry, Report (1977-1982) F.T. 6/9/1978; Chemistry and Industry 3/12/1977;"OECD Interfutures: Nutrition and Agriculture Report. [June 1979].

2. See OECD: "Report on the State of the Environment" and OECD - Interfutures: "Long term perspectives of the world car Industries" - [June 1979].

3. See : CEE, L'Avenir, de l'Industrie Automobile dans la Communauté, Dec. 1976.

a fair increase in mobility and car population, emissions are likely to increase considerably; only strengthening the emission standards would limit this trend.

On the other hand, account must be taken of the present tendency to promote public transportation, in particular in dense urban areas. This tends to generate beneficial environmental impact in terms of reduced air pollution, congestion and in terms of resource consumption such as energy and land. Noise might also decrease provided public transport vehicles, such as buses, are significantly quietened.

However, the positive impact of new standards and replacement of old by new cars is limited for by four factors:

- reductions in effective emission levels by an average vehicle were particularly important for CO, but are lower for HC; for NO_x emissions control proved particularly difficult and has not been successful;
- the efficiency of emission control systems degrade significantly over time and it has been estimated that, for a 5-year old vehicle, CO and HC emission can be twice as high as the required standard;[1]
- reductions in vehicle emissions have primarily concerned automobiles; not much has been done in the case of motorcycles, so that pollution from those sources, which is a important part of vehicle pollution would not be significantly reduced by the stock replacement effect;
- reductions in the noise emitted by motor vehicles have not been impressive, and the stock replacement effect will certainly not compensate for the stock increase and mobility increase effects; in the U. K. for instance, it is expected that the number of city dwellers exposed to high traffic noise will increase by 600,000 over the 1978-1985 period (i. e. about 15-20 per cent increase); significant increases of noise exposure are expected in other countries.[2]

Aircraft noise is also a major problem and the progressive replacement of old aircraft by less noisy types constitutes a favourable trend, significant reduction of exposure to aircraft noise cannot be expected before the late 1980s. Hence the need for insulating dwellings around airports.

It seems fair to conclude that the environmental implications of likely trends in automotive transportation are serious if 1978 standards

1. See "Cost and effectiveness of automotive exhause emission control regulations" - OECD, 1978.

2. See:"Reducing Noise in OECD Countries", OECD, 1978; and "Report on the State of the Environment - Part II "Environmental Conditions,"("high" noise means more than 67 dBA leq).

and policies are not strengthened: in particular, increased pollution and increased noise.

1.4. Urban Problems

The sectors discussed in this report are, to a significant degree, closely linked to urban problems. Cities are centres of industry, transport, energy consumption, and also, to an extent, tourism. They expand at a cost to agricultural land and natural amenities. Any shifts in urban development trends are therefore bound to have - via accompanying changes in the location, intensity, and technology of economic activity activity - important consequences for the environment. Conversely, changes in the value attached to environmental quality, and amenities in general, will in turn influence the locational and investment decisions of households and firms.

Recent urban trends are beginning to expose the need for a better appreciation of these relationships between the economy and the environment in an urban context. The 1970s have witnessed the consolidation, and in some countries the acceleration, of three aspects of urban change, each of which is closely tied to environmental concerns:

i) Urban decline has taken place in the older industrial regions of all OECD countries; as the result the traditional economic base has been lost to other regions or countries. The environmental impact on formerly dominant inner cities has been severe: a lack of maintenance of infrastructure and housing, a declining tax base from which to finance improvements, vandalism and crime accompanying economic and societal decay. The problem of environmental decay is, in such cities, replacing previous problems of environmental pressure: congestion, air pollution, noise. The result is the creation of large tracts of urban land whose characteristics are an incentive to the further flight of investment capital and further decline.

   The continuation of structural change in OECD economies, on a scale that affects large areas of cities, is expected to continue through the 1980s. There are in fact indications in North America that the problems of environmental and economic decline are beginning to spread to post-war suburbs.

ii) Accelerated urban growth in environmentally-attractive areas has become the counterpart of urban decline. The source of this growth is two-fold: economic activity moving away from deteriorating areas, and totally new growth taking place in

regions which, often more for their environmental advantages than for traditional locational advantages, are considered to be the growth centres of the future.
The result has been the rapid increase in the problems of urban growth, including environmental costs, in regions lacking a tradition for financing, legislating or regulating that growth. The enabling factors behind this trend - a shift to higher technology, less resource-dependent industry, improved transport and communications, and higher personal incomes - will continue to cause sufficient urban decentralisation in the 1980s that environmental policy will be continually faced with the need to adapt to shifts in locational preferences and reduce the environmental costs of growth without detracting from growth potential itself.

iii) The less developed economies in OECD countries, especially in the Mediterranean area, are now experiencing rapid urban growth. Both forces driving this trend will continue through the next decade at least: closer international economic interdependence in trade and investment, and increasing discretionary income being spent on tourism from more developed countries. The next decade will, without the introduction of land management strategies and increased expenditures on behalf of the environment, be one of serious environmental degradation in some previously depressed or relatively underdeveloped regions. This will, unless anticipatory policies and programmes are introduced quickly, create the urban conditions which may lead to the kind of eventual economic decline now being experienced in some older economies.

In sum, urban problems are intricately related to environmental concerns in general, and to economic development trends in particular. Many of the environmental problems examined by OECD in recent years reach their peak concentrations in cities, and their solution will continue to depend on government action in response to urban economic growth and change.

1.5. Energy

Energy is a major source of pollution at the production, transportation and consumption stages; in most countries, it is the single largest discharger of SO_2, CO, NO_x and dust. The implications of expected developments in energy on environmental quality therefore deserve attention. The two most important considerations relate to consumption and structural changes.

Energy consumption is clearly associated with economic growth. For many years, until 1973, energy consumption increased at about the same rate as GDP, 5 per cent per year. In other words the ratio of growth in energy consumption to growth in domestic product in OECD countries was about one. In the period that followed, marked by significant increases in energy prices, this ratio decreased dramatically to 0.5. In some countries such as the U.K., over the 1972-75 period, an increase in GDP was even accompanied by a decrease in energy consumption. There is some uncertainty as to the value that this ratio will assume in the medium term. International Energy Agency forecasts put it at 0.84. This would mean an increase of 19 per cent in energy consumption over the 1978-1985 period, with a 3 per cent per year growth rate of GDP, and an increase of 30 per cent with a 4.5 per cent growth rate of GDP.

In most OECD countries[1] major efforts to develop viable replacements for imported oil, such as domestic offshore oil and gas, coal and nuclear power, have been initiated. Although those efforts are not expected to change significantly the shares of the various sources of energy, except for nuclear energy, they should lead to considerable increases in absolute terms. Solid fuel production in OECD countries is projected to increase at a 3.8 per cent rate over the 1974-1985 period, and domestic oil production at a 3.3 - 4.3 per cent rate over the same period. This means that the potential pollution emissions as well as the environmental and health hazards, associated with such production could increase by about 30 per cent between 1978 and 1985.[2] In addition, electricity production is likely to increase at much higher rates; very large modern power plants, whether fossil-fueled or nuclear-fueled, as well as transmission power lines, can also have disruptive impacts on communities and landscapes and can for, social and other reasons, meet with increasing opposition.

It is in the field of energy conservation that policy goals coincide and are attainable simultaneously. Indeed experience shows that they mutually depend on each other. Pre-1974, environment concerns alone were insufficient to generate conservation policies; post-1974, environmental support for, and environmental benefits from, conservation were needed to sustain and justify conservation policies improvements. Even without changes in 'life-style' in all sectors - household, transport and industry - there is a substantial potential for energy conservation, and the reduction in emission and other environmental impacts would be at least proportional.

For Denmark it has been estimated that, even without adding the environmental benefits, the cost of additional investment for conservation of about DKr 20-billion would be offset by an accumulated savings

---

1. "Report on the State of the Environment", Part I.

2. An OECD scenario puts the likely increase in SO_2 emissions for the period 1974-85 at about 24 per cent; "Clean Fuel Supply", 1978, OECD.

of DKr 30–40 billion in imported fuels between 1975-1990. Swedish energy policy has a clear objective to reduce energy demand from a growth rate of 4.5 per cent to 2 per cent on the average between 1975 and 1985.[1]

Finland has achieved considerable savings in energy economy through the use of district heating (two thirds of the housing stock) and co-generation of heat and electricity in industry; these energy measures were accompanied by a reduction in emissions and consequently improvements in air quality. A financial scheme for energy conservation measures has been highly successful in Norway; the cost of investment per unit of energy saved was found to be substantially lower than the cost of new energy produced.

If these conservation efforts were sufficiently wide-spread they could reduce significantly the growth of emissions from energy production and use. At present, however only a few OECD countries have vigorous and comprehensive energy conservation programmes in place. Others have taken strong conservation measures in some sectors and still others have prepared legislation aiming at improving programmes. A half dozen countries have not yet undertaken major energy conservation initiatives.[2]

1.6. Tourism

One of the activities in the tertiary sector with significantly growing environmental impact is tourism.

The environmental concern about tourism stems from its fast growth both at the international and domestic level, and from its unexpected environmental consequences. Tourism has increased rapidly over the last two decades and it is expected to continue to increase rapidly in the coming years. It has been growing well above the rate of GNP growth. For example, between 1945-1977 international tourism grew from 25 million arrivals to 240 million. The forecast for 1990 is 580 million. In the North-West zone of the Mediterranean (Italy, France, Spain) the tourist population could grow from 52 million in 1973 to about 85 million in 1985 or at about 5 per cent per annum. This expansion will take place on an essentially fixed ecological base. Consequently the beach area in square meters available per tourist will shrink by about 50 per cent.[3]

The economic benefits of tourism are undeniable. Receipts from international tourism grew from $5 billion in 1960 to over $40 billion in 1977. It produced important foreign exchange earnings, over half

1. "Environment and Energy Use in Urban Areas", OECD, 1978.

2. "Energy Policies and Programmes of IEA countries, 1977 Review," OECD, 1978 and "1978 Review", OECD, 1979.

3. "Environment and Tourism", OECD, 1980.

of the import bill in some countries. It was also used for regional development and was an important source of employment; major tourist-receiving countries employing 5 per cent of the work force in tourism related activities.

Uncontrolled tourism however can damage the environment. It brings noise, water pollution, solid wastes, aesthetic destruction, soil erosion, and loss of green and agricultural areas. It often intrudes in previously unpolluted areas, such as coastal resorts, rural areas and moutain zones, including ecologically highly sensitive areas.

This environmental degradation is bad in itself. But is also represents a critical threat to tourism. Tourism which has played an important and useful role in the development of some countries or areas, and should continue to do so, could become "self-destroying": if it degrades its most important attraction, a good environment, the growth of tourism could fall to an insufficient rate.

Given that tourism is likely to continue its rapid growth it is essential that tourist and environmental authorities co-ordinate their activities to maintain the environmental quality of tourist areas. In their co-ordination they have to take steps to ensure that at least the normal environmental standards such as for air, water and noise are maintained. In addition stricter than normal standards migh be needed for traffic and building density and for the safeguard of green areas. To ensure even the minimum requirements two steps are essential in the coming period[1]:

i) to clean up the back-log of pollution in its many varied forms and provide the appropriate infrastructure to deal with pollution;
ii) to introduce the appropriate legal, industrial and planning mechanism for the future orderly development of tourism in line with environmental needs.

1.7. Summary Comments

The environmental consequences of projected economic developments in the coming years will not be negligible. How significant they will be depend on a number of factors.

Comparison with the past decade could be misleading; in that period, most countries enjoyed high growth rates. Because of vigorous environmental policies, they also secured certain environmental improvements. It is possible, however, that in the future even with lower economic growth there will be some degree of environmental degradation.

---

1. "Draft Recommendation on Environment and Tourism;" [C(79)51].

In the absence of improvements in pollution control, associated in most cases with tighter environmental policies, a significant deterioration of environmental quality could occur over the period to 1985.

In a number of cases and pollutants, attempts were made to estimated increases in pollution emissions that could take place between 1978 and 1985: in most sectors the figures could be as high as 30%. This is a very rough order of magnitude of potential growth in emissions associated with projected economic growth, present emission standards and constant technology. These calculations disregard the accumulation of non-degradable pollutants in the environment.

If, however, present environmental standards were relaxed, deterioration would be greater. On the other hand stronger policies would seem to be needed to maintain or improve environmental quality in the future.

2

CONSTRAINTS AND CHALLENGES FOR ENVIRONMENTAL POLICIES

The introductory section to this report, in presenting possible outcomes for world economic developments, pointed out that over the foreseeable future policy makers will be faced with difficult choices and constraints on their room for manœuvre. This section presents a brief review of the possible issues which could arise in this area.

2.1. Employment

In the present economic situation, unemployment is a major concern in many Member countries. Not only does unemployment exceed 5 per cent in several Member countries, but also the employment situation is unlikely to improve during the early 1980s (see Introduction). Unacceptably high unemployment is a fact that must be taken into account in formulating environmental policies during the medium term. Specifically, high unemployment can affect environmental programmes in two ways. First, high unemployment is sometimes thought to justify the relaxation or postponement of discharge standards in order to avoid plant closings and output reductions that might increase unemployment. Second, high unemployment is thought to justify government environment expenditures, perhaps financed by government deficits, to stimulate employment. Both beliefs require careful analysis.

The strongest fears concern the possibility of plant closings because of environmental controls. Plant closings resulting at least in part from pollution control requirements have been reported in several Member countries[1], but the evidence shows that most plant closings were of marginal plants, which would have closed relatively soon even in the absence of pollution control[2]. In addition, closings had impact on a relatively small number of jobs.

1. See "Employment and the Environment" - OECD 1978 - In the United States, the EPA "Early warning system" has indicated for the period April 1976-March 1977 - 15 plant closures affecting 2, 546 employees raising the total for 1971 through March 1977, to 107 plant closures affecting 20, 318 employees The Council on Environmental Quality - Eighth report - 1977 .

2. In the United States, a survey of the Department of Commerce (1974) reported that in 78 per cent of the cases where pollution control was allegedly a cause of plant closure, decreasing sales, increasing operating cost, obsolescence and similar factors in fact contributed to the closing.

In a period of deep concern with high unemployment levels, the problem of plant closings should not be understated; however, it must be put in perspective i. e. as a short term problem of minor scale.

Recent evidence shows that environmental policies can create jobs. In the United States it was estimated that about one million persons were employed in jobs related to pollution control activities in 1975 (total expenditure $ 15.7 billion); it is projected that in 1980 the additional expenditure of 4.7 billion dollars will create 193,000 jobs. In Germany, environmental policy has created about 218,000 jobs during the period 1970-1974 and is estimated to create 366,000 other jobs for the period 1975-1979.[1] In France, the manufacture and operation of pollution control equipment employed 173,500 persons in 1978, including 44,500 involved in waste disposal.[2] Comparison of various estimates show that environmental programmes create between 40,000 to 67,000 jobs per billion dollars of expenditure. Macro-economic studies indicate that environmental programmes bring, in the short term, a global net creation of jobs. In the longer run, employment effects could be either slightly negative or slightly positive.[1]

There is no consensus in OECD countries on the desirability of using environmental programmes for job-creation. However it is recognised that Government environmental expenditures do create jobs and have multiplier effects in much the same way as do government expenditures for other purposes. Consequently some countries are using such policies for that purpose under special conditions; Norway, Denmark and Sweden have used environmental programmes for job-creation; in Sweden, environmental programmes were expanded during 1971-74 and this created about 21,000 jobs. The present reflationary and job creation programme in Finland has also a strong environmental content.

Environmental programmes are candidates for increased government expenditures, along with other expenditure categories, when governments want to stimulate employment by increased expenditure. It is generally accepted in modern macro-economics that government programmes to put the unemployed to work have little or no opportunity cost. But it does not follow that it makes no difference at what activity

---

1. "Employment and Environment" op. cit. see also "Macro-Economic Evaluation of Environmental Programmes" - OECD - 1978. This finding is supported by the Norwegian study showing that the expansionary impact on employment, produced by pollution control requirements, is more than twice the adverse effect due to price increases and reduced profits.

2. "Environment et Reorientation de la Croissance" (Ministère de l'Environnement et du Cadre de la Vie - Septembre 1978).

the unemployed are put to work and how the programme is financed. Among programmes with a given opportunity cost, benefits may vary greatly. Thus, the basic role should be that government programmes to stimulate employment should increase employment in activities whose benefits are greatest. Thus availability of unemployed resources does not avoid the need to calculate programme benefits.

If environmental programmes are to stimulate employment, they must be carefully designed. Many construction projects, e.g. of waste treatment plants, take many years to plan and execute, and may be of little help with short term, cyclical employment problems. By contrast, some programmes can be affected quickly. Programmes for home insulation and neighbourhood rehabilitation appear to fall in this category. Similarly, a programme to stimulate construction activity may be of little value if the construction industry is operating near capacity and unemployment is concentrated in other industries. Finally, unemployment is often concentrated in certain depressed regions; if so, expenditures that stimulate employment elsewhere may be of little value for employment in the region concerned.

2.2. Inflation

Inflation has been at the forefront of public concern over the last few years and will presumably remain so in the foreseeable future, particularly if growth rates and employment accelerate again. Faster growth, by allowing more rapid increases in real income may diminish wage pressures, but it could revive demand pull forces in both labour and product markets. Commodity prices, in particular, could respond rapidly to a revival of demand in OECD countries as they did during the short-lived stock-building boom of 1976.

Clearly then the impact of environmental policies upon prices and the level of inflation is an issue of importance. The Polluter-Pays Principle, adopted by OECD Member countries, is designed to ensure that the cost of environmental measures is reflected in the costs of goods and services. If this cost is passed forward in price rises to the final consumers the change in relative prices will induce consumers to shift their purchases towards less polluting alternatives. In this sense price changes associated with many environmental measures constitute appropriate economic responses. It should be recognised that the Polluter Pays Principle aims at a greater efficiency in resource allocation; indeed the PPP is defined as an economic efficiency principle[1], and the increased efficiency should have an overall favourable effect on the price structure.

The impact of environmental policies upon national rates of inflation is likely to be quite small in comparison to much more basic

---

1. See the 'Polluter Pays Principle', OECD, 1975.

causes of the present high inflation rates. This is a finding supported by studies that have been made for the United States, Japan, the Netherlands and Norway[1], which estimate the direct and indirect effects on prices of pollution control measures so far taken. At the overall level of the consumer price index, for the United States at the end of the 1970-77 period the index was 2.3 per cent higher than it would otherwise have been (i.e. the inflation rate was increased by an average of about 0.3 per cent per annum). For Norway it was estimated that the full 'passing-on' of the cost increase in industry the consumer price index for the 1974-77 period would rise by 0.4 per cent. For Japan the rise in the price level was estimated for the five years 1972-77 to have been increased by a total of between 2.2 and 3.8 per cent (i.e. between 0.4 and 0.7 per cent per annum).

For individual sectors the picture is more varied, with price rises particularly noticeable in branches such as pulp and paper, basic chemicals, iron and steel and automobiles. Overall, however, it does not seem that the environmental measures will make more than a marginal contribution to inflation: a small price for the improvement in the quality of life and the economic benefits of pollution control.

2.3. Public expenditure

Government environmental programmes are of three kinds: regulation of pollution discharges, direct provision of environmental protection services and research, development, monitoring and data collection. In most Member countries this second category comprises the bulk of government environment expenditures.

Gross government expenditures are a large share (between 30 per cent and 40 per cent) of total environmental protection expenditures. Often, water pollution control (especially construction and operation of sewage treatment plants) accounts for a large part of government environment expenditures in Member countries (about 30 per cent). Expenditures for air pollution control invariably fall most heavily on the private sector. There is no analog in air pollution control to government collection and treatment of wastes in water pollution control.

Government environment expenditures are a large part of total environment expenditures in most Member countries. At the same time, however, environment expenditures are a small part of government expenditures in most countries : usually less than 1 per cent. This percentage reflects the fact that total (public and private)

---

1. "Macro-economic Evaluation of Environmental Programmes', op. cit. and "Price Effects of Norway's Environmental Programmes, " (unpublished).

environment expenditures are still only a small part of GDP in Member countries: mostly under 2 per cent.

i) Future trends in Government Expenditures

The postwar period has been a time of rapidly growing government expenditures in Member countries. Total government expenditures in OECD Member countries increased from 28.5% of GDP in 1955-57 to 41.4 per cent in 1974-76.[1] Thus, Member countries environmental programmes have been put in place during a period of not only rapid overall economic growth but also of a rapidly growing government share of GDP.

Questions are now being raised in many Member countries as to whether further growth in government's share of GDP is desirable. In some countries, strong pressures are emerging to stabilize and even reduce the government share of GDP. Whether pressures result in such a reduction or not environmental protection programmes could face increased competition for government funds during the 1980s.

What programme categories are likely to provide the strongest competition for government funds? The first is transfer payments and income maintenance expenditures. During the sixties and early seventies, half of the total increase in government expenditures was for "welfare", including "quality of life", programmes. In the future, strong pressure may be felt for increases in income maintenance programmes such as unemployment compensation and old age pensions. Health care expenditures are also likely to continue to rise rapidly; but environmental protection to a certain extent could be seen as a preventive health care measure.

The second, is the increasing demand for employment subsidies. They may take the form of direct subsidies to alleviate the burden of rationalising industrial relocation, (e.g. because of foreign competition). Alternatively, they may take the form of increasing government employment as a means of reducing unemployment levels.

The third, are the growing claims on government funds for direct assistance to industry for restructuring and rationalisation necessitated by changing demand patterns and increasing international competition.

The fourth is related to the increase in the share of the welfare programmes which was facilitated by the relative decrease in defence expenditure that occurred during the last 10 years or so in some Member countries. It is now likely that this decline in the share of defense will be halted in a number of countries.

---

1. "Public Expenditure Trends" - OECD, 1978.

ii) Future Trends in Financing Environmental Programmes

The period to 1985 is almost certain to be one in which new ideas and efforts are needed to ensure the financing of environmental programmes. Slow economic growth will make both government and private financing difficult. Public resistance to further growth in the government shares of GDP will limit availability of government funds, and competition for government funds will be intense from other expenditure programmes.

One implication of this stringency is to place greater emphasis on cost-benefit and cost-effectiveness analysis. Cost-benefit analysis can help to calculate justifiable total expenditures. More important, it can guide funds and resources to the most productive programmes, so that scarce funds are better used.

One solution to the dual problem of efficiency and financing environment expenditures during the 1980's is to place more emphasis on the "polluter-pays principle".[1] The PPP demands that producers pay the full costs of pollution abatement. The result is that pollution abatement costs are likely to be incorporated in product prices (whenever the market structure allows), enabling consumers to take into account the full social costs of their purchasing decisions. Governments should ensure the implementation of the PPP.

But more could be done in combination with direct regulations (e. g. emission standards). Emissions or effluent charges could further improve the ability of consumers to take account of the social costs of production, could reduce emissions, and could provide revenues to governments to cover the cost of environmental programmes.[2] Some Member countries already make use of pollution charges for these purposes.[2] Likewise user fees could be employed to finance larger share of the costs of sewage treatment plants. User fees not only reduce the burden of such costs on government budgets, they also ensure a better balance between the use of government-provided treatment plants in contrast with self-treatment and product or process changes. They also tend to be more easily accepted by the public as they establish a direct link between the payment and the corresponding benefit (purchase of a service). User fees also introduce a market discipline into the allocation of public resources.[3]

---

1. See "The Polluter Pays Principle" OECD, 1975.

2. See "Pollution Charges", OECD - 1976. The Recommendation on "Noise Abatement Policies" [C(78)73 Final] and the Recommendation on "Water Management Policies and Instruments" [C (78)4 Final].

3. See "Public Expenditure Trends" op. cit.

To conclude, the fact that public environmental expenditures are a small part of overall public expenditures makes them vulnerable but not a major constraint on other competing claims. In other words, if public expenditures on environment decreased, expenditures in other areas could gain little. Environment, however, would lose much.

2.4. Investment, Profitability and Productivity

A crucial requirement needed to achieve a desirable recovery in economic activity in most Member countries is a return to much higher rates of investment. A number of countries witnessed some deceleration in investment starting in the early 1970s, but it has been particularly low in the recent past for obvious cyclical reasons.

While market forces may be relied upon to achieve some increase in investment, particularly in bottleneck sectors or in energy production, the rather pronounced shift towards investment implicit in the higher growth scenario may call for government action. In the short-run, action may be necessary to increase demand (and thereby the utilisation rate of capacity as well as productivity). But over the longer-run a sustainable increase in investment may require a more permanent increase in profitability and reduction in the uncertainty and risks surrounding investment decisions and the financing of business activities.

At a global level, recovery of investment would mean a quicker incorporation, given existing regulations and standards, of pollution control equipment of the "built-in" rather than of the "added-on" type in the capital stock. Indeed this could even reverse the recent declining trend in environmental investment in those countries which had introduced vigorous programmes in the early 1970s.

As for the impact of environmental investment two aspects should be considered, namely:

i) Like most other investment, it could contribute to maintain the growth of aggregate demand at a time at which this may be a government priority;

ii) It could, however, have an impact on the measured growth of productivity from the perspective of the firm and this in turn could adversely affect the return on investment (though it may well encourage innovation and, of course, it improves welfare in a large number of areas).

There are four possible areas where global investment and environmental investment could interact. They are: profitability, uncertainty, financing and productivity.

In so far as profitability is viewed as being crucial to a return to high growth rates, there is the possibility of an inconsistency between the requirements of macroeconomic and environmental policies. It is in no way apparent, on the basis of available statistical evidence, that environmental measures have contributed noticeably to the already mentioned declining trends in profitability at the aggregate level. (These, in most cases, started well before pollution control measures were introduced). At the sectoral level some negative effects may have taken place in recent years, though evidence suggests that they have been relatively small in relation to other problems such as low levels of overall demand, rising wage or material input costs, or increasing foreign competition.

In relation to uncertainty connected with environmental investment, governments could provide notice well in advance of introducing regulations and give a sufficiently long-period of transition to counteract the claim that environmental regulations introduce additional risk to investment. Indeed the PPP caters, to an extent, for this particular eventuality. Anticipatory policies, looking at medium-term problems would be particularly helpful to mitigate such an impact.

Rigidities in the financial markets, such as strict profit criteria for loans could also retard environmental investment. Earlier OECD exminations of financial assistance schemes have shown that the raising of funds for environmental purposes from the market could be difficult and more costly than for other investment, particularly for medium and small businesses. Additionally recovery of investment activity could, by putting some financial markets under pressure, accentuate these difficulties.

The rate of productivity growth in recent years has been low as compared to past growth rates and environmental policies may face pressures insofar as they are identified as contributing to this fall in productivity.[1] The fall in productivity is likely to reflect a number of factors e.g. reduced possibilities for improved efficiency arising from economies of scale in industry and agriculture, a slower rate of major technological innovation, and particularly, at present, low rates of capacity utilisation.[2]

There is a tendency to underestimate the contribution of environmental expenditures to productivity for several reasons:

---

1. e.g. E.F. Denison, in an article 'Effects of Selected Changes on the Institutional and Human Environment upon Output per unit of Input', published in the Survey of Current Business, Janaury 1978, suggests that expenditures to protect the environment have contributed to lower productivity, as shown in the narrow accounting sense of measured output per unit of input.

2. Economic 'Outlook' OECD, July 1977.

1) Conventional measurement, using as a definition of productivity the ratio output-over-input (capital and labour input), does not count as output the improvements to the environmental and to general welfare achieved by environment policies. Consequently the measurements ignore the benefits of environmental measures in terms of better health, reduced loss of man-hours better attention to work and reduced losses in transport.

2) In national accounting procedure private expenditure on pollution control is normally counted as an intermediate not as a final expenditure.

3) The accounting approach assumes away the evidence that new equipment with in-built environmental control adds to output over and above what the old equipment without environmental investment would have produced. Indeed there is evidence that environmental regulations have contributed to the generation of important new innovations both in processes and products (see Part 3).

There remains a clear need for measures of national welfare in addition to the classical output measures such as GNP, which include estimates of improvements in welfare associated with environmental changes. For the cases where the benefits of environmental programmes have been estimated (see Part 3) their contribution to real growth of national welfare appears sizeable.

2.5. Environmental Policies and the Balance of Payments

Theoretically environmental policies could have some effect on international trade of goods, on the flow of capital and raise trade v barriers to the free flow of international trade. Therefore in this context the following aspects are examined:

- the impact of environmental regulations on the cost of internationally traded goods; the impact on the balance of trade of internationally traded environmental capital equipment;
- the impact on the flow of capital between countries with significantly different environmental standards and investment in polluting industries;
- the impact of environmental regulations as non-tariff barriers to trade.

i) Environmental Cost and Internationally Traded Goods

The polluter-Pays-Principle was adopted by OECD Member countries a) to encourage rational use of environmental resources and b) to avoid distortions in international trade and investment. It was also agreed that governments should seek harmonisation of environmental policies and should not assist polluters in bearing the cost of pollution control. Furthermore, it was recommended: "Where products are traded internationally and where there could be significant obstacles to trade, Governments should seek common standards for polluting products and agree on the timing and general scope of regulations for particular products".[1]

In following up these recommendations the Environment Committee examined the extent and the impact of financial assistance systems provided for environmental purposes and concluded: "The examination indicated that significant distortions in trade probably would not occur at the aggregate level. However, because the examination was carried out at the aggregate level, no conclusions could be drawn at this stage at the sectoral or industry level."[2]

At the same time it was noted that emission and ambient standards and the timetable for their achievements vary considerably amongst Member countries and these differences are reflected in trade advantages to industries in certain countries.

On the basis of available evidence it can be provisionally concluded that even with existing differences in standards the overall trade effects of environmental policies appear on the whole either neutral or they work in such a way as to reduce the possible trade imbalances between OECD countries.

The problem is somewhat different when individual industries are looked at and competition from non-OECD countries is brought into the picture, particularly for those industries with high level of idle capacity.

ii) Capital and Investment Flows

In spite of repeated claims that environmental requirements have induced significant investment shifts to countries with low levels of environmental control, no such evidence has been put forward. Studies lead to the conclusion, that other factors e.g. wage and material costs, vicinity of markets, security of capital, exchange rate movements and other factors are far more important than cost differentials due to environmental controls.[3] The claim that environmental programmes have led to the creation of pollution havens has not, so far, been substantiated. For example the net formation of new foreign affiliates of U.S. companies has been declining for some years in response to the decline of relative costs per unit of industrial output in the U.S., this has occurred in spite of the relative stringency of U.S. environmental regulations.

---

1. "The Polluter Pays Principle" OECD, 1975.

2. Results of the Procedure for Notification of Financial Assistance Systems for Pollution Prevention or Control, OECD, C(77)129.

3. Macro-economic Evaluation of Environmental Programmes OECD, 1978.

iii) Environmental Regulations as Non-Tariff Barriers to Trade

In certain cases involving noxious products which are dangerous to the environment, environmental regulations would mean the prohibition of use of the product; this may be the case with chemical agents such as biocides. The exports of such products from the producing countries would be affected by the prohibition of imports into another country. Such a policy would be neutral between different manufacturers of the same product but would give preference to the producers of less noxious substitutes.

In spite of the rapid growth of non-tariff trade barriers in recent years only in a few of these cases has environmental protection, been used as the justification for them. Indeed there have been international efforts to obtain agreement on the environmental impact of such noxious products and to harmonize the introduction of these regulations and thereby limit their possible trade impacts. The on-going discussion on toxic chemicals within the Chemicals Group of the Environment Committee aims directly at minimising or eliminating the possible harmful trade effects of the environmental regulations on toxic chemicals.

Looking at the Balance of Payments effects of environmental policies as a whole the available evidence suggests that they were minor and insignificant relative to the upheavals which occurred in recent years. OECD countries have made a major effort at the international level to minimise the potential impacts. The harmonisation of environmental policies is contributing to the containment of these effects.

The relationship with non-OECD countries has become more important since 1974. It is a two way impact; some 'OECD' environmental policies have an influence on the economy of non-OECD countries; in turn the regulations in those countries, or the lack of them, could change trade patterns. These relationships are likely to be more important in the future. For example the potential shift of hazardous and high-pollution industries to countries in the Third World may loom large in the 1980s. Such shifts have been threatened although evidence that any has occurred primarily to escape the impact of anti-pollution legislation in an OECD country, is hard to come by.

Summary Comments

The economic constraints that are likely to prevail over the period up to 1985, although real, do not pose a major conflict between economic and environmental objectives and policies to achieve them. They call, however for judicious selection of environmental policies and they are likely to enforce an economic discipline in environmental management to obtain the best value for money. They also show the need for assessing the significant benefits of environmental improvements, especially in the field of productivity, in terms which allow comparisons with the benefits of other policies.

3

PROSPECTS FOR ENVIRONMENTAL POLICIES

3.1. The Present Challenge

The preceding sections have established that future developments in OECD countries have certain implications both for the state of the environment and for the strength of environmental policy. The magnitude and the structure of future growth create a potential for a significant increase in the pollution load by 1985. This is the case even though the expected economic growth rate will be lower than in prior periods.

This lower rate of real growth, however, may make it more difficult to maintain and enforce existing environmental policies, and may impose constraints on efforts to improve the effectiveness of environmental measures. With lower productivity contributing to both inflation and unemployment problems, it may be argued that diversion of private investment to environmental capital projects must be reduced. The long-term decrease in enterprise profitability can appear to reinforce this argument as do assertions regarding some adverse employment and price impact of environmental policies. Also governments confront both financing constraints and increased demands from competing programmes.

Clearly a major challenge in the 1980s will be to find the right balance between quantitative and qualitative growth and the choice will be influenced by the following considerations :

i) the demand for basic human needs, food, shelter, clothing etc, are being increasingly met in the majority of OECD countries;

ii) consequently the environmental side-effects of quantitative economic growth are perceived as having more pervasive consequences on the welfare of the individual.

Within the economic framework of this paper two types of implications emerge for environmental policy making. One is the efficiency of environmental policies, which could be improved through the increasing use of economic incentives in addition to regulations. More recent innovations in this field include the emission offset market which would require firms to obtain emission reductions from other firms in a region before adding to it.

The second is the need to measure the magnitude of the benefits to be gained from environmental measures.

Some estimates of national benefits and costs of pollution abatement have been calculated but they have not been developed to the point at which it is possible to locate the expenditure level which equate incremental benefits and cost. The available national studies, however, suggest that total benefits are well in excess of total costs of pollution abatement.[1] It is likely that, in most countries, incremental costs and benefits will be equated only at a higher environmental quality than has yet been achieved. Recent estimates suggest that, whereas pollution abatement expenditures in Member countries amount to 1-2 per cent of GDP, pollution damages range between 3 and 5 per cent of GDP.[2] Although the benefit figures are rough estimates, they suggest that, in economic terms, the present situation entails a welfare loss and that pollution abatement expenditures should be increased.

3.2. The Benefits of Environmental Improvement

The continued strong support for environmental improvements is notable, given the difficulties of measuring and quantifying the benefits of quality improvements. The costs of achieving environmental improvement are visible and, indeed, recorded in both national income accounts and estimates of economic productivity changes. Although it is widely acknowledged that GDP has its limits as a welfare indicator and that maximising GDP does not maximise welfare, this fact is rarely taken into account in policy discussions.

It should be emphasised that while relativities might change, the basic justification for pollution control programmes remains the same in good times and bad, when prices go up or down, when unemployment is high or low: a high quality and low-pollution environment improves the welfare of and is valued by citizens. Air pollution is a hazard to health, health, property and crops. Water pollution entails health, recreational and aesthetic damages. Noise becomes an omnipresent nuisance affecting health and welfare of a growing proportion of populations. Pollution control expenditures are justified by the benefits they produce, even if these benefits are difficult to state in monetary terms.

---

1. a) A recent U.S. study calculated that benefits from a 60% reduction in air pollution of SO_2 and particulates would result in a $ 43 billion (1978 prices) worth of annual benefits from reduced mortality and morbidity. b) The latest data for 1977 estimates that cost of incremental pollution abatement expenditures for air at $ 12.2 billion. Sources: Unpublished Report of Contractors to the Environmental Agency, U.S. "Methods Development for Assessing Trade-Offs in Environmental Management" 1978 and The Ninth Annual Report of the Council on Environmental Quality.

2. "Environnement et Réorientation de la Croissance" op.cit.

The benefits of pollution control are difficult to measure because the output, a high quality environment, cannot be bought and sold in markets. As a result, indirect measurement techniques must be employed. Such techniques are inevitably approximate and controversial. But progress has been made during the last decade. There are official and private estimates of health damages from air and water pollution and on the impact of noise. There are now also many estimates of recreational damages caused by water pollution. Because air and water pollution abatement have some characteristics of luxury goods, available evidence suggests that the benefits of a high quality environment rise rapidly as income and leisure time increase.

In addition to the benefits listed earlier in this section it should be emphasised that the welfare gains yielded by pollution controls also produce indirect and long-term economic benefits: savings in health expenditures, improved labour productivity in a less-polluted and quieter environment, improved productivity of land if crops are not damaged by air pollution, and so on. A high quality environment can appropriately be viewed as both an intermediate good which contributes to production and a final good demanded in itself. And, finally, environmental programmes can be used to create both economic demands and employment in a slack economy.[1] In Japan, for instance, the production of pollution control equipment reached 693.9 billion yen in 1976, as compared to 34.1 billion yen in 1966. By 1976, the number of employees in industries manufacturing pollution control equipment and facilities reached 50,000. In the U.S., markets for pollution control equipment and products in 1977 totalled roughly $ 1.8 billion and are projected to grow to a level of $ 3.5 billion by 1983.[1]

A further indirect benefit from environmental policies has been termed "technology stimulation", i.e. environmental policies have induced innovations in production processes. Indeed, regulation perhaps more than anything else, has forced firms to engage in wholesale rethinking of their production processes and products. Often this shocks them out of a rather inflexible production system and thereby provides the catalyst which is necessary for innovation to occur. This is most likely to occur in mature and highly concentrated industries where the pattern of innovation is characterized more by process changes than by new products and where market conditions pose few challenges to the status quo.

Such "technology stimulation" has been documented in the U.S. When the initial requirements of the Clean Air and Water Acts began to take effect, many industrial firms reevaluated their entire production

1. For instance, in France, value added by pollution control activities represented 0.5 per cent of GDP and 0.7 per cent of exports (1978). Source: Ministère de la Qualité de la Vie, Paris.

system in terms of product recovery and waste minimization. The process review led many firms to adopt innovations that led to significant economic benefits, including energy savings.

In Japan also the setting of stringent standards led to rapid technological innovations in comparison to other countries. When PCBs were banned, substitute products were developed at low additional cost. When strict standards on automobile emissions were set, several ways of meeting them were found and, in contradiction to the extravagant cost estimates which had been associated with these standards, no significant increases were recorded. The Japanese experience in the field of pollution abatement lends support to the idea that to a large extent it is not technology that constrains policy choices but policy choices constrain technology.[1]

In Norway stringent air pollution control measures triggered technological innovations in the production of ferrosilicons by forcing a change from open to closed furnace operations. This resulted in energy savings, better working conditions and reduced air pollution. The savings in the cost of production varied between 8 - 12 per cent per unit of production improving the profitability of the operation.[2]

According to a major study[3] at the Massachussetts Institute of Technology, this technology stimulation is wide-spread among OECD countries. In examining the relationship between environmental/ safety regulation and innovation in five major industries in France, Germany, Holland, the United Kingdom and Japan, the study found that environmental regulation may change the course of innovation and result in more successful innovation in certain industries.

Examples of enterprises that actually decreased production costs as a result of pollution abatement measures are given in a study prepared by the French Ministry of the Environment.[4] Cost reduction is achieved through two main mechanisms: the necessity to re-think and re-design production processes leads to technological improvements, and the discovery that some previously unrecovered pollutants have economic value (SO_2 contains sulphur, and the dust emitted by a cement factory consists largely of cement).

---

1. "Environmental Policies in Japan" 1977, OECD.

2. Paper presented at ILAFA International Congress on Technology, May 1978, Acapulco Mexico, by Elkem-Spieger Verket, Norway.

3. National Support for Science and Technology: "An examination of Experience in Five Countries," Center for Policy Alternatives Massachusetts Institute of Technology. Cambridge, Mass. U.S.A. May, 1976.

4. Ministère de la Qualité de la Vie - Environnement, Usines Propres: La Technologie au Service de l'Environnement, Paris, la Documentation française, 1977, 58 p.

Public satisfaction with the environmental achievements is evidenced by the continuing strong demand for environmental improvements, and, in particular, opposition to major new sources of environmental disruption. For example, in the United States, recent opinion surveys indicate that environmental quality is an important concern, and ranks at about the same level as concern with health and education. In 1977, 47 per cent of the people surveyed believed that insufficient financial resources were allocated to environmental improvements. Another survey in 1978 in the U.S. showed that the public wants both more energy and environmental protection, however, 47 per cent of those interviewed preferred environmental protection while only 31 per cent preferred energy.[2] Environmental concerns such as clean air and clean water rank as high or higher than other basic concerns to Canadians, such as quality of food, housing, education and social services. More than half of Canadians expressed the view that they would pay higher taxes if they felt the money was going to anti-pollution measures and more than three-quarters said they would pay more for some products on the same basis.[3] Similarly, in France, surveys show that environmental degradation is perceived as one of the most important problems.[4] However, where the goal of environmental protection is compared with other socio-economic objectives (such as unemployment and inflation) public support for environmental improvement is somewhat less strong. In France, pollution control and nature conservancy ranked just behind employment but before inflation in 1976.[4] A 1977 survey rank industrial production (i.e. employment) before pollution abatement, whereas 29 per cent would give priority to the environment.[1]

These patterns vary among countries depending upon living conditions and the economic situation. Neverthess, it appears that the demand for environmental protection continues to be strong, though challenged by growing concern with macro-economic performance. It should be noted, however, that the demand for environmental protection is very responsive to increases in income; there is a high income elasticity. Even if income grows more slowly than in the 1960s, the demand for a higher quality environment will expand. Both output increases and environmental quality improvements are given high priority and are in the public mind not in conflict.

---

1. "Environment an enduring concern" - Resources for the Future - January/March 1978.

2. "The Public speaks again: A new environmental survey" - Resources for the future - September/November 1978.

3. CROP - Public Opinion Study, 1978.

4. S. Antoine and J. Navarin, Les Français et la Qualité de la Vie", Futuribles No. 14, Mars/Avril 1978.

3.3. Environmental Policies for Future Economic Conditions

The likely economic growth patterns for the medium term require both changes in existing environmental and economic policies and development of new strategies for achieving environmental improvements. While the specific design of policies will vary according to the economic and the environmental conditions prevailing among countries, two general points can be made. First, environmental policies should be oriented to improve their efficiency. Second, environmental policies and policies with other goals should be integrated so as to have the greatest possible environmental benefits while minimising the overall social costs.

i) Reorientation of Pollution Control Policies

The policy measures undertaken in past decades have had beneficial impacts. Some indicators of total pollution loads have turned down in spite of significant increases in production. And, industry-specific pollution-output ratios, which were thought to be fixed, have fallen in many sectors. Real welfare benefits have resulted from these changes. Although private and public financial costs of abatement activities appear large, the net costs are not, and usually negative when all the benefits are properly accounted for. In economics characterised by high levels of unemployment, as OECD economies have been since 1975, some portion of the labour and capital devoted to environmental protection would otherwise have been unemployed. In some cases, both public and private pollution abatement expenditures are useful additions to aggregate demand, as has been recognised by some Member countries (e.g. Sweden, Denmark and Norway) that have used such measures as anti-recession policy instruments.

These recorded gains, however, say little, about the economic efficiency of the policy instruments which have been chosen. Indeed, it is possible that significantly greater improvements in environmental indicators could have been obtained with the same budgetary allocations if more efficient policy measures had been implemented. Several suggestions for the future orientation of environment programmes can be offered.

First, PPP should be applied consistently to induce a shift wherever possible in relative prices in favour of low pollution industries. This would induce producers to pass on the full costs of meeting environmental standards, resulting in the inclusion of pollution abatement costs in product prices and enabling consumers to take into account the full social costs of their purchasing decisions. Through this approach, government subsidies and expenditures could be reduced as well.

Second, emission charges and user fees would improve the ability of the public to take account of environmental damages caused by production, would result in further pollution abatement and would provide the necessary funds to finance environmental public expenditures. Appropriate user fees would enable those who generate wastes to make rational cost-minimising choices between use of government sewage treatment plants and alternative abatement procedures such as self-treatment and process or product charges. Most important, user fees provide funds to finance at least part of construction and operation of treatment plants, thus removing part of such expenditures from government budgets. [1]

Third, it is now recognised that in several instances, the design of regulation procedures lead to some efficiency losses. Individual bargaining between government enforcement officials and polluters may sometimes lead to unnecessary transaction costs and result in inefficiency and inequity. A shift towards increased use of incentive systems in which polluters would find it in their interest to reduce emissions and away from direct and costly case-by-case rule enforcement would be more cost effective. User and effluent charge procedures, are consistent with this shift towards an incentive (not subsidies) strategy and their application together with market approaches should be evaluated parallel with regulations. The incentive system should take account also of the variability of local absorption capacity of the environment together with the use of variable charges.

Fourth, it has been argued that technological improvements in pollution control are one of the key factors in reaching present and future, possibly more strict, environmental standards. The past has shown a relatively rapid rate of progress in technological improvements; in order to keep up this progress, it is suggested, that given the potential return at the national level governments should consider supporting financially, as taken into account under the PPP, research and development in the field of pollution abatament technology. Apart from new abatement technologies environmental R and D should also aim at identification of up to now undiscovered or new sources of pollution and new or better monitoring systems.

Finally, governments should increase their focus on the efficiency with which pollution abatement equipment is employed. Until now, the main policy concern has been to produce and install pollution abatement equipment. Even with the application of the best available technology approach not enough attention has been paid to the efficiency of its operation. Inefficiencies in operation stemming from excessive breakdown times, uncorrected equipment failures, tardy repairs and untrained personnel are now well documented. Relative to the past decade, the next years should find relatively more resources allocated

---

1. See "Pollution Charges" op. cit and "Recommendation of the Council on Water Management Policies and Instruments" [C(78)4-Final].

to the improved operation of installed equipment rather than the installation of new equipment. The design of policy measures which neither subsidise nor excessively regulate private and public activities is a difficult matter, to which substantial attention should be given.

ii) Integration of Environmental and Other Policies

In addition to redirecting and streamlining the operation of existing pollution control policies, programmes with other goals can be reformulated to have beneficial environmental impacts. This normally requires coordination between environmental and other agencies.

1. Intensive use of energy and raw materials is accompanied by potentially high pollution loads. Hence, gains in environmental quality could be achieved by co-ordinating energy and raw materials policy with environmental policy. Such policy co-ordination would entail the recovery of raw materials waste and use of waste heat; lengthening the product lives; and substituting low "throughput" processes or products for those which are high users of energy and raw materials.

The choice among various energy sources (and to a lesser extent between various types of raw materials) is also relevant. While such choices are often not decided on environmental grounds, they have environmental implications. Policy co-ordination should ensure that the environmental costs of alternative choices among energy and raw material sources are reflected in decisions.

2. In the field of employment policies, decisions could be adapted to serve environmental purposes. As mentioned above, environmental policies and expenditures can have the incidental effect to create jobs and thus contribute to employment objectives. The issue here is to implement employment policies so as to contribute to environmental quality objectives. At present, public service employment is often proposed for both cyclical and structural reasons. Given the range of environmental problems, especially in urban areas, it seems that some of the jobs created by these programmes could be designed so as to achieve environmental objectives. Another major employment policy on the horizon is the possible use of wage (or employment) subsidies to stimulate private sector job creation. If used such subsidies could be structured so as to encourage both population and industry dispersion from congested, high pollution areas.

3. In most countries, environmental deterioration is more severe in large, congested urban areas. Gains in overall environmental quality could be achieved by dispersion of high pollution activities to sites with lower population concentrations and with better assimilative capacity. But dispersion policies should be formulated taking account of the advantages of location in metropolitan areas in particular their economic costs.

4. A common objection to modern production facilities is their enormous size. Because economies of scale may not be as important some new technologies as in those used in the past, the development of smaller industrial complexes with a proportionately smaller overall environmental impact can be encouraged. This is consistent with the encouragement of regional dispersion noted in 3) above.

5. Another area where environmental objectives can be achieved through policy co-ordination is in the area of housing and urban development policy. Urban policies, i. e. policies developed to make cities more pleasant and more efficient in providing resources, can be given a strong environmental component. It is not enough to aim at pollution-free cities; one must aim also for amenity-rich cities. Amenity production is an important environmental objective, and it is largely a matter of urban design and control. It includes the preservation and renovation of old houses and buildings, as opposed to tearing down and reconstruction. Renovation is not necessarily more costly and its resource content is different: it utilises more labour, less energy, fewer raw materials and less capital.

6. "Time-planning" policies are related to location or "space-planning" policies. Both aim at reducing congestion. By reducing the number of people or activities which seek to use the same area of resource, such policies can both reduce congestion and improve environmental quality. Better arrangements of work and leisure over the day (peak hours), the week (weekends) and the year (holidays) could contribute to both objectives.

CONCLUSIONS

The report explored in detail the likely environmental developments, in particular the potential growth of pollution and policies to deal with them during the coming period up to 1985 in the context of a feasible economic scenario. The significance of this economic approach lies in the fact that economic factors have been in the past and will be in the future one of the dominant determinants of environmental needs and demand and as well as of environmental policies themselves. The growth and level of per capita income in OECD are closely related to the ability and willingness of countries to spend on the environment. The level of income, in turn, is closely related to economic growth which itself is the generator of potential pollution and other environmental effects.

The conclusion that emerges from the paper is that because of this close relation between environment and economic development the changed economic circumstances and trends to 1985 give rise to concern and provide a new and possibly difficult set of conditions for environment policy.

Even though economic growth will be slower the environmental consequences will not be negligible. Assuming present emission standards and technology, potential emissions in a number of sectors could rise 30 per cent by 1985. In addition the greater reliance on indigeneous energy resources could mean a similar increase in emissions from energy production and use. These emissions are additive to the already existing levels in the various regions and will call for stricter emission standards even to maintain present ambient quality. These could in turn lead to more costly emission controls at the margin. There are in addition a host of new pollutants which are only now receiving close attention for their possible harmful impacts and a list of new sources of pollution which will have to be dealt with in the coming years.

Over the last 10 years or so, during the first phase of major environmental programmes the OECD economies were able to absorb easily the cost of pollution control. In a number of countries these programmes even added to economic growth and to employment without any very significant inflationary impact. In the changed economic conditions greater emphasis will have to be placed on the complementarity and compatibility of environmental and economic policies. The impact of environmental policies on productivity and subsequently profitability will be important in a period when investments and productivity are crucial to an acceptable rate of economic growth. The possible impacts of environmental policies on inflation, employment and balance of payments will have to be taken into account in devising policies.

Under these conditions new environmental policies are likely to be subject to increasing scrutiny not only for their economic and social impact, but also for their efficiency and cost-effectiveness. Decision making will be facilitated if the benefits of environmental measures, which proved to be substantial and well exceeded their cost in the past, are enumerated and evaluated in meaningful and where possible, economic terms.

The re-orientation of environmental policies towards more efficient prevention, control and conservation can be partly achieved through a greater use of economic incentives. The most significant contribution to efficiency in the past came through technological innovation generated by economic forces, both in environmentally-favourable production processes and in control techniques. They are likely to be the most important factors in the future in reducing control cost.

To anticipate successfully environmental problems or to pre-empt them from occuring, the integration of environmental and other policies in a number of fields is a major prerequisite and the main policy strategy. Some success through environmental impact assessments have been achieved but substantially more is needed in the field of industrial policy, urban planning, land-use planning, transport, tourism and energy policy. It will be a major challenge for environmental policy making to achieve a successful integration with other policies both during policy formulation as well as in implementation. Without such integration anticipatory policies are likely to be a failure.

The picture provided by this report could have been softened perhaps had it been possible to examine in detail and generalise certain structural changes starting to emerge locally in a limited number of OECD countries. These changes are the first signs of a transformation from the industrial era to what is popularly described 'post-industrial era'. This transformation is a very gradual one and its possible environmental implications have not been explored. One of the most dramatic demonstrations of this transformation appears in the information-electronic revolution with its low raw material and energy requirements. The converse development is the relative decline in the basic heavy industries. It was not possible to discern a general major trend in OECD countries in this direction for the period under discussion but it should be an aim of future policies to detect and prepare for these gradual changes even if their impacts are likely to be felt only in the more distant future.

IV

HOW TO INCLUDE ENVIRONMENTAL CONCERNS IN THE DECISION-MAKING PROCESS

SUMMARY AND CONCLUSIONS

The present study takes French experience[1] as its point of departure, adjusted in the light of what the author knows of experience abroad. It will therefore appear incomplete, since it does not reflect the problems of all OECD countries. Nevertheless, France seems to be one of the countries which has encountered the widest range of difficulties and has, therefore, had to deploy the greatest variety of solutions.

It seems to the author that the fields of activity in which the greatest efforts should be made in future to bring environment considerations to bear as an integral part of the decision-making process are:

- energy (nuclear power stations, high-voltage power supply systems). Other countries may have specific problems, such as in regard to oil or mining plants;
- urbanisation and tourist development;
- polluting industries and installations (including livestock facilities);
- transport, in terms both of infrastructure (motorways, airports) and modes (motor vehicles, aircraft);
- chemical products dispersed into the environment, and diffuse sources of pollution.

The principal ways in which to make decision-makers in these fields aware of the need to include environmental considerations as one of the factors in the decision-making process would appear to be:

- training, mainly by fostering awareness of the issue at initial training level;
- information, through media appropriately chosen to reach the decision-maker, through competent protection associations and

---

1. This paper has been prepared by Mr. J.F. Saglio, former Délégué à la Qualité de la Vie in the French Ministry of the Environment and Quality of Life.

by providing the general public with clear, straightforward information about the state of the environment;
- improving the quality of environmental administrations;
- modifying economic concepts which are detrimental to the protection of the environment;
- research on environmental protection methods, and the publication of economic and technical findings

There would appear to be three broad lines along which to proceed in order to avoid major conflicts between economic development and environmental protection:

- information at both national and local levels, appropriately tailored to the public it needs to reach with respect to message and media;
- genuine participation by intermediate bodies in the preparation of decisions;
- compensation for damage inflicted as a result of decisions.

There are in fact many procedures and mechanisms with which to incorporate environmental protection 'upstream" in the decision-making process. All may be used concurrently to the extent that each is more or less adapted to a particular type of problem and to different cultural traditions and administrative practices. Individual countries may have national, regional or local environmental protection systems and bodies in a very wide variety of forms.

Anticipatory mechanisms such as explicit, publicised, strictly enforced legislation and regulations, procedures for public consultation, economic mechanisms (fiscal, grants, finance) complying with the Polluter-Pays Principle, the internalisation of costs, optimum use of national assets (land-use plans and quality objectives), environmental impact studies, all have their use and may be applied simultaneously. Employed in combination, they make it possible to bring concern for the environment to bear upon the decision process at the earliest possible stage, a factor promoting efficiency and savings.

I. INTRODUCTION: A NEW PERSPECTIVE

1. World economic changes during the past four years have led authorities in many countries to raise the question of environmental protection. Opinion surveys continue to show that the three main sides to this issue - the struggle against environmental degradation, protection of animal and plant life and improvement of the quality of life - are subjects of major public concern for which strong policy action is demanded, which continue to plague certain Member countries. Nevertheless, in general today, an increased importance is attached to the economic consequences of proposed environmental protection measures. At the same time at the local level strong opposition is raised against the creation of economically sound projects for reasons of environmental protection.

2. These difficulties show that we have yet to acquire the habit of systematically dealing with environmental values, on a footing with economic or technical factors, as evident components of decision-making. Hence a relatively new question which arises is : "how in the present and future context can environmental concerns be included in the decision-making process?". In fact the problem now is to weigh environmental concerns both with respect to those who would ignore such concerns or consider them incompatible with present economic requirements; and to those who would exaggerate their importance.

3. It is always more economical and more expedient to introduce environmental concerns as promptly as possible into the decision-making process. [1] Three examples taken from the French experience serve to illustrate this premise:

- it is more difficult and costly to treat emissions from an existing industrial facility than to introduce purification facilities in designing a new plant, to achieve equally effective pollution control. This has been verified in particular through pollution control programmes undertaken since 1972 in certain existing industries (pulp and paper, sugar refineries, cement works, yeast manufactures, etc.);
- in Western France a vegetable canning plant was set up next to a small stream with extensive spawning grounds for salmon despite advice to the contrary from environmental authorities. In fact, only a few kilometres away an equivalent site was available on the banks of a larger and less vulnerable stream capable of tolerating eventual fluctuations in the water treatment system.

---

1. There are many decisions affecting the environment in this report. However, only investment decisions taken by authorities other than those directly concerned with environmental protection are considered.

Trusting existing technology, the manufacturer maintained his initial site selection. In fact fluctuations did occur, and the manufacturer will now have to spend far more on additional treatment measures and equipment than would have been the case had the location been shifted in the first place;

- studies have shown that in suburban areas the additional cost of new mortoways designed to minimise the inconvenience to nearby residents from noise is approximately 2 to 4 per cent. But it would cost twice as much to obtain equivalent results for existing motorways in similar areas.

4. One claim often heard, however, is that taking account of environmental considerations from the start leads to undue expenditures and delays. Yet experience has shown that if environmental factors were appropriately examined and included at each stage of the decision-making process, delays and extra costs were generally reduced to a minimum.

5. Questions which seem relevant in this context are the following:

a) What are the difficulties to overcome in order to ensure a better integration of environmental concerns in the decision-making process?
b) What are the types of situations and areas of activity most likely to lead to major conflicts between development and environment?
c) What are the mechanisms and institutions for and at what level (national, regional, local) should the effort be made?

II. DIFFICULTIES TO OVERCOME IN INCLUDING ENVIRONMENTAL CONSIDERATIONS IN THE DECISION-MAKING PROCESS

6. Experience has shown a number of examples of such difficulties: the inadequacy of traditional economic accounting, ignorance, misgivings about the economics of the project, employment concerns, competitiveness, fears of delays and misgivings with respect to environmental protection authorities and the "fait-accompli".

The inadequacy of traditional economic accounting

7. At the present time no mechanism exists to measure in economic terms improvements in general well-being resulting from measures of the environment. In fact, Gross National Product is the sum of the value of industrial production and the cost of controlling pollution and nuisances, whether this cost arises from expenditures for prevention or expenditures related to redress of damages caused by pollution.

The search for an increase in GNP does not encourage a reduction in the sum of these two activities and, therefore, does not encourage actions to prevent environmental deterioration. This constitutes a substantial distortion factor. The logical situation would be to develop another type of national accounting and steps in this direction have been taken in the United States and in Japan.

8. A second factor of distortion is the absence of the notion of "environmental assets" in national accounting systems. It is thus possible to have excellent operating accounts, even if such assets are depleted. Few countries have undertaken to apply such an accounting system. None use it in connection with natural assets and environment as a whole. This second distortion factor should be corrected. The two above-mentioned changes will not immediately solve environmental conflicts but will lay the ground for a consistent conceptual approach as between the economy and the environment.

Ignorance

9. The decision-maker in good faith is often unaware of the concerns to be taken into account and the directives to be observed. It is therefore necessary:

- that government bodies be vigilant and that their sources of information be adequate;
- that the decision-maker realises that he must approach the proper authorities responsible for the preparation of protection measures;
- that the decision-maker be aware of the main steps to be taken. These will be determined through contacts with the appropriate governement agency.

Misgivings as to the economics of the project

10. The decision-maker worries about the investment (and operating) costs of environmental protection installations or measures which may threaten the economics or profitability of his project. This assumption is currently popular. The decision-maker's only regard and sole concern is for his own direct interests, not those of the community. It is indeed not uncommon for locally elected officials to take up and advance this argument.

11. Experience shows that while environmental constraints may have a considerable impact in the modification of an ongoing project, they very seldom have any decisive effect as to the project actually being carried out. A project with exceptional effects on the environment

may however be rejected (such as an open-cast mine or an airport).

Misgivings about employment

12. A frequent effect of the preceding case is the feeling that employment may be threatened by economic imbalances brought about by environmental constraints. Experience so far, however, fails to show whether these actually do influence the level of employment, which may be affected locally or in a particular economic sector but not on a nation-wide scale nor as regards the overall economy.

13. It would therefore seem advisable to circulate widely any studies or assessments which may have been made regarding the effect of environmental constraints on employment.

Competitiveness

14. Can pollution controls or environmental protection measures lead, as claimed by certain industrial decision-makers, to distortions in international trade with respect to various products, or can it lead to the establishment of polluting factories in countries with less strict standards?

15. In most cases this fear would appear to be groundless and cost-price distortions due to other factors would seem to be much greater, e.g. overall labour costs, costs of supplies, location of markets, political stability. Moreover the Polluter-Pays Principle is such as to reduce these distortions.

16. Undue differences in pollution control regulations between countries should be avoided. One useful approach would be to make periodic comparisons of these regulations.

Misgivings in regard to delays and mistrust in relations with environmental protection authorities

17. One fear of the decision-maker is that discussions with environmental protection authorities for defining the precautions required in the decision-making process may lead to costly measure or undue delays. This argument is frequently advanced, especially in the United States, by manufacturers who see such an essential dialogue as a factor contributing to a slowing down of productivity gains. But the experience of environmental authorities is that protracted negotiations do not lead to stricter directives and instead result in fewer constraints for the industrial decision-maker.

18. It is therefore appropriate to ensure that government departments can function as smoothly as possible by:

- establishing general and public rules covering the processing of dossiers, including the setting of time-limits;
- providing especially careful training to environmental protection authorities; continuing training should be organised or improved;
- adapting administrative resources to the workload;
- setting time-limits for processing and creating meachanisms for seeking recourse in the Courts for compensation if such time-limits are exceeded.

The "Fait-accompli"

19. This attitude is one particularly adopted in the case of important public decisions. The public authority dislikes reversing decisions, notably for political reasons (examples are motorway layouts). In order to avoid reversing a decision, fallacious economic reasons may be cited. Alternatives are presented as unduly expensive, while all or part of the costs to the community resulting from the contested decision are usually forgotten.

20. Such problems could be avoided if environmental protection considerations were taken into account sufficiently ahead of time in the decision-making process. Once again, this implies a vigilant, sufficiently staffed department of the environment which is independent enough to oppose public projects and firm enough to impose environmental concerns at the right time.

Conclusions on means to overcome the above-mentioned difficulties

21. The main causes underlying the above-described attitudes and difficulties are the lack of training, the inadequate provision of information, operation of the administrative agency, the inappropriateness of classical economic concepts, and the inadequacy of economic and technical data. Suggested solutions should be aimed at:

- developing a general educational programme on the environment for schools and universities rather than just for specialists;
- helping to set up environmental protection agencies and strengthening their capacity to influence the decision-making process;
- regularly publishing reports on the state of the environment as well as on the reasons for choices made in important decisions affecting the environment;
- adjusting the operation of the government bodies (for example, through training, simple publication of regulations, means to avoid delays, better co-ordination, etc.); in any case it would

appear that a specialised agency with overall responsibility for environmental protection is necessary, even if it be part of a larger ministry responsible for other sectors;
- modifying national accounts (resource accounts, taking account of the social costs of activities);
- promoting research on environmental protection methods and the extensive publication of economic and technical findings.

III. ACTIVITY AREAS AND SITUATIONS LIKELY TO LEAD TO MAJOR CONFLICTS BETWEEN ECONOMIC DEVELOPMENT AND ENVIRONMENT

22. The means available to environmental protection authorities are limited. They must therefore focus their efforts on those areas and in situations where their means will have the most effect. To this end many OECD countries have set up special committees to review projects in such areas.

Areas for priority action

23. These areas share the following three characteristics:

- they involve heavy potential risks and serious consequences for the environment (natural environment, animal and plant-life, landscape, quality of life);
- whether stable or developing, they affect an important or increasing proportion of the natural environment, country or population;
- they involve issues about which the public is especially sensitive, whether in terms of the size of the populations concerned or the of its reactions.

24. On the basis of complaint statistics, risk analysis and technical and economic development projections, the following may be regarded as priority areas:

25. Energy, in particular nuclear power stations, high voltage power supply systems and installations for reprocessing irradiated fuels and for storing nuclear wastes. These installations share certain characteristics which especially lend themselves to environmental protection efforts:

- they are in the development stage, take up space and alter sites, particularly in the vicinity of seas and streams in the case of power stations;
- the size of the units is considerable and is likely to increase;

- the potential risks attributed to such stations or waste treatment facilities and the actual effects on the landscape of these facilities and high voltage power supply installations are acutely perceived by the public.

26. Urbanisation and tourist development. Urban development as well as that of tourist facilities and holiday homes appear to be priority areas for concern from an environmental standpoint. In fact, this is an area of growing concern. It is a source of serious conflict in the matter of land use, alterations of the landscape and physical planning. The consequences for living conditions and the natural environment are significant.

27. Polluting industries and installations. While this is a traditional environmental concern, it is perhaps now more than ever a source of conflict, especially at the local level between environmental protection on the one hand, and, on the other, promotion of the economy and employment.

28. Transport. Transport infrastructures (roads, motorways, airports) and transport facilities (automobiles, motor-cycles, trucks, buses and aircraft) constitute an essential area for environmental protection. Their effects on the environment include noise, altering the landscape and modifying living conditions.

29. Special reference should be made to oil tankers and vessels threatening massive accidental pollution: public feeling towards such disasters runs extremely high and the political consequences may be serious.

30. Chemical products. The public is becoming increasingly sensitive to the dispersion of chemicals in the environment, whether the natural environment or man himself is affected.

Conflicting situations

31. At the heart of these priority areas various situations are likely to lead to major conflicts between development and the environment. These situations fall into three main groups, namely:

- issues about which the public is particularly sensitive;
- cases in which the aims of the enterprises are at odds with those of the community;
- cases of unresolved conflict between local, regional and national interests, one party overriding the other without compensation.

32. Various examples may serve to illustrate these types of situations:

Special public sensitivity

33. A number of chemical plants near the mouth of the Seine: mostly refineries, fertilizer plants producing phosphoric acid and discharging gypsum and titanium oxide, plants which discharge acid wastes.

34. Estuary pollution is a sensitive subject. The discharges by the Montedison company into the Mediterranean have further sensitized the public about titanium oxide plants dumping wastes into the sea or into an estuary.

35. A good deal of effort has been made to control industrial waste pollution of the Lower Seine. A gypsum recovery plant, making plaster panels has been established; and gypsum wastes have been ferried out to sea areas chosen by fishermen themselves as being least harmful to aquatic life. Discharges from the titanium plant have been reduced to a technical and economic minimum (and one plant has even been closed down). The refineries have equipped themselves with the best available waste purification system.

36. But despite all this, the local residents, fishermen and seamen, are still dissatisfied; the latter may perhaps be blaming pollution for economic difficulties which are actually due to other causes (equipment and fuel costs, obsolete boats, over-fishing, etc.).

37. The problem has given rise to a number of demonstrations and lawsuits. A settlement now seems to be in sight, following the establishment of commissions on which representatives of the fishermen and the public are associated with monitoring of the estuary and discharge control and also following the award of various forms of aid to the fishermen.

38. In the Etang de Berre and at Fos, in the South of France, there are heavy industrial concentrations, mainly refineries and petrochemical plants. The prospect of large-scale industrial development at Fos, involving steel, smelting, energy and chemicals, aroused strong local public feeling. It is a relevant point here that the fishermen using the Etang de Berre had already received considerable compensation through regaining their fishing rights when a dam was built to alter the water characteristics of the Etang de Berre.

39. A Commission made up of government officials, elected representatives, fishermen and private local residents was established six years ago, before the planned Fos development. Its role was to monitor

discharges and to be consulted in the setting of discharge levels. This commission has worked well. Conflicts have been forestalled ahead of time. The region is now much cleaner, albeit industrially more developed.

Aims of the enterprise differing from those of the community

40. Publicly-controlled or private firms may have distinctly different aims from those of the community in which they are located, while nonetheless being profitable to the national or overall economy.

41. The creation of land-fill sites, whether publicly or privately controlled,increasingly meets with opposition from representatives of the local communities concerned. Whatever the local residents may be offered as compensation and however carefully they may be assured as to the quality of the operation, generally speaking, public consent is not reached.

Unresolved conflicts between local, regional and national interests

42. Reservoir dams may be taken as an example of this situation: France is developing a programme of dams on the Loire, to control flooding and increase the rate of flow when the river is at its lowest. The social value of this programme is beyond dispute. But there is lively local opposition which has been amplified by various movements which associates this programme with secondary aims dealing with more controversial subjects such as, cooling for nuclear power stations. One source of such opposition seems to be inadequate compensation. Obviously, lands to be flood are purchased and material losses are compensated, but there is no compensation for damages caused by an alteration of the countryside. Furthermore, local residents have the feeling that a source of collective wealth is being created in which they will have no share. It is noteworthy that a number of local politicians are setting themselves up as "pure-water champions".

43. The example above is an instance of a project which is of value to the country as a whole, but involves an unsettled local conflict. The converse frequently occurs; for example, when a ski resort is planned.

44. A snow-covered site, suitable slopes, good road access, a favourable exposure, all these may encourage the local community to set up a ski resort. The national community, through the government, may oppose this for various reasons: perhaps for economically based reasons, for example if the project would compete with an existing resort in the vicinity, jeopardising the development of the latter and

making its own viability doubtful; or there may be reasons of an ecological nature, e. g. if the project would spoil an outstanding natural area. The conflict between the local community, which favours development of the project and the State which opposes it, is understandable. But in a case like this, it would be very hard to imagine any sort of mechanism by which the government could provide compensation for a project which has not been implemented.

What kind of solutions can be envisaged?

45. Three possible approaches would seem to emerge: information, participation and compensation:

- local or national information is often the way to clarify a situation, revise defective projects, forestall unfeasible projects, explain the aims in view and secure their acceptance:
- participation, not only by the public, but also and in particular by intermediary bodies (such as trade associations, chambers of commerce, etc.), and by local-elected officials, is essential. Participation should mean that responsibility is conferred, i. e. that responsible individuals are identified and express themselves, and foremost among these should be elected representatives;
- compensation for damage which leads to rational economic choices is a necessity. The consensus required for the collective optimisation of decisions will no longer be attainable unless some compensation (monetary or otherwise) is provided for the various forms of direct or indirect damages; or even unless there is some redistribution of the collective wealth created by a public infrastructure.

IV. INSTITUTIONAL INSTRUMENTS AND MEANS OF INTEGRATION

46. There are a number of administrative and political frameworks as well as mechanisms and devices enabling environmental concerns to be included in decision-making. These largely depend on the organisational traditions and practices of different countries. The level at which decisions are taken, i. e. national, regional or local is also a factor. In theory, the more important the decision the higher the level at which it is taken, but this is not necessarily true for those decisions which have an effect on the environment: thus the building of a house near a unique monument may destroy the site. Even though the decision is local, the authority responsible for protecting the site may be national. Hence, while there is an obvious advantage in taking environmental

concerns into account at the same level as the decision-making authority, they may also have to be considered at a higher level.

Mechanisms and instruments

47. These may be of different types: legislative, regulatory, economic:

- Laws and regulatory mechanisms

Among the most recent laws adopted in certain OECD Member countries, those aimed at the expanded control of chemicals prior to their marketing and use are typical of anticipatory legislation.

A marked, anticipatory effect is obtained, if the laws and regulations are explicit, publicised and strictly enforced.

- Economic mechanisms (financial incentives and penalties, government subsidies, access to special funding and local community aid). Experience demonstrates that economic incentives on even a modest scale which comply with the Polluter-Pays Principle can have a very favourable effect on decisions where the environment is concerned.

48. Four mechanisms of particular importance are:

- land-use plans
- charges
- environmental impact studies
- quality objectives.

Land-use plans

49. Land-use planning was introduced in France thirty years ago. Its aim is to make the best use of this limited collective asset to reduce direct and indirect collective capital and operational expenditure, to improve the quality of life (land-scapes, way of life, etc.), permit consistent, harmonious building development, protect activities (agriculture) etc. Land use permits are finalised after agreement between government and elected representatives, following public consultation.

50. Experience over this 30-year period indicates the following advantages and disadvantages:

- the zoning of land parcels confers different potential values upon them and creates undeserved capital gain for some land-downers. A building site is worth more than land which cannot be built upon. It is very difficult to establish equitable compromises between owners;

- land-use plans cannot be sacrosanct, but must evolve with growth. It is not easy to achieve the proper balance between the necessary flexibility and the degree of permanence required to ensure effectiveness of the plan;
- finally it is difficult to achieve agreement between divergent interests of central government (reduction of public expenditure, protection of unspoiled landscapes), local elected representatives (urban development), and landowners.

51. This mechanism proves extremely effective for land use, therefore for improving the quality of life, but is of little help in protecting the natural environment (because it makes no provision for authorising development, nor for standards to be observed in the use of land). Some countries (e. g. the United Kingdom) however carry such planning further and in fact do prescribe how the natural environment is to be used.

52. Various improvements might be possible:

- greater public participation: for example, by publishing zoning ordinances in the form of booklets for mass circulation;
- less cumbersome procedures in certain cases (small communes);
- national directives imposing constraints in the form of general regulations for particular zones (coastline, mountain areas);
- in countries where the administrative organisation so allows, uses of the land and natural environment could be planned to a greater extent.

Charges

53. France has for ten years had a cost internalisation mechanism, in the form of Basin Agencies[1]. These impose charges on consumers or polluters of water in proportion to the amount of water drawn, consumed or polluted. The money thus raised is used to subsidise equipment for improving water resources, quantitatively or qualitatively.

54. Experience shows that these charges act as incentives to reduce pollution and consumption and to encourage operational efficiency through anti-pollution equipment, the selection of less polluting or less heavily consuming techniques, to improved plant siting of installations. Nonetheless, complementary regulatory action would appear essential to encourage clean-up, in particular in those instances where the charges are modest.

---

1. Agences de Bassin.

55. Similar mechanisms are found elsewhere, as in the Netherlands for water use and in Japan for air pollution. Proof of their effectiveness is shown, but also that regulatory action remains an indispensable need.

The environmental impact study

56. Various forms of this mechanism exist in several countries, including France since 1st January, 1978 and the United States and Canada since 1970. It is applied essentially to both public and private investment decisions requiring government authorisation. This is the main "environmental audit" process for incestments; its effectiveness in decision-making is just beginning to be demonstrated, and it appears to be the most promising.

57. In France, those projects to which this mechanism has been applied had naturally been ready before the procedure came into operation. Its effects on decision-making are therefore still limited, leading mainly to modifications in certain aspects of a project without changing the basic principle. But impact studies are influencing the very design of new projects. This has also been observed in the United States and Canada, where major projects, even when begun, have been abandoned.

58. When this procedure is fully implemented one inconvenient effect it may have is to delay decision-making. This is due to two reasons:

- the procedure is being applied to projects which have already been designed. However, in the future, it will be applied "upstream" of the decision-making process and should not continue to be a cause of delay;
- the methodology has not yet been settled. It is not yet known what main points should be brought out by the impact study according to the type of project and where it is to be located.

Experience in the United States and Canada shows that these shortcomings disappear over time.

59. An advantage of this procedure is that, if applied as early as possible in decision-making, defective projects are rejected at the outset, while still at the preliminary study stage. This is a definite advantage, although no figures can be provided since the decision to cancel is taken by the originators of the project before any public investigation.

60. The cost of impact studies has been criticised. In the United States it varies considerably (between 0. 05 per cent and 1 to 2 per cent of the value of the project) but remains acceptable. The few available French statistics show a similar order of magnitude.

61. Necessary improvements would appear to be as follows:

- the impact study should be prepared at the earliest stages in decision-making, examining suitable elements in increasing detail as the formulation of the project proceeds;
- reliable technical and economic data should be available, again showing the importance of research;
- according to the type of project, and where it is to be located, the elements requiring study should be distinguished from those which are secondary;
- genuine alternatives should be considered: this implies that a project forming part of a larger whole (high tension power lines, power stations, roads, motorways), should be studied before component parts are considered;
- methods of public consultation should be adapted to the type of project.

62. It may be asked who should be responsible for carrying out the impact study: the originator of the project, the branch of government which controls it or the branch of government concerned with environmental protection? The present author regards this as a somewhat secondary question, and feels that each country should find an answer in line with its own traditions. In France, responsibility lies with the project originator. The main point is that the study should be publicised, and that the public should know what action has been taken on the findings.

Quality objectives

63. This mechanism implies the exploitation of a natural asset in a coherent geographical context: for example, the use of water in a basin or river. Taking account of the actual and foreseen uses, of the level of pollution of the resource, of its limitations, of investment and operating costs imposed by the struggle for pollution abatement or by the increase in resources, the administration draws up several schemes for utilisation of the water courses. They attach to these schemes the advantages and liabilities of each and outline the necessary work plans. Representative users, publicly consulted, select one of these schemes. Abstraction and discharge permits are ultimately delivered in compliance with this scheme. Difficulties are similar to those encountered in land-use planning. This mechanism of setting quality objectives is particularly efficient for anticipatory purposes in decisions concerning very fragile or over-used water basins.

Institutional means

64. The following are examples of such structures which have already been set up in different countries:

Interministerial meetings at national level

65. This mechanism is used in two types of cases:

- when the problem addressed is still new;
- and/or when the problem addressed is complex and calls into play the authority of several different ministerial departments or government agencies.

66. The advantage of this framework for action is to bring together government authorities at the level necessary to prepare overall national policy, but suffers from the disadvantage of the absence of political power and the need to work for unanimity.

Minister and government department

67. A Minister's responsibility may be of two different types depending on whether he is concerned exclusively with the environment or simultaneously with the environment and other areas. The single-jurisdiction ministerial structure is preferable when the problem dealt with comes only under the authority of the Minister and when he has a department exclusively or primarily responsible for carrying out his decisions. On the other hand, when a Minister has other responsibilities, there is a risk that the other areas of concern will prevail over the environment or vice-versa. Nonetheless, two considerable advantages emerge: the policy-making power of such a Minister is greater and, environmental protection hence further ensured under his personal responsibility.

68. It is important to have a government department whose task is to protect the environment. This is not to say that all relevant functions of the government should be concentrated in one department, but in it there should be a clearly identified focal point responsible for environmental protection, if possible with regulatory powers and authority over public funds.

Planning bodies

69. The bodies (commissions) responsible for preparing plans have a most valuable part to play in anticipating decisions, this moreover being their essential role.

70. They do not generally have political power, or the authority to decide directly, and their work is in fact a compromise among their members. But that work takes place within a consistent framework, where the principal factors are taken into account, which is a great advantage.

National public institutions

71. There are a number of instances in France: the Basin Agencies, the Higher Fisheries Council, the National Hunting Council and the National Waste Recovery and Disposal Agency.

72. These are extensions of central government, responsible for improving the environment in some limited sector (water, wildlife, solid waste, etc.). Their resources are dependent upon how effectively they operate, which is a powerful spur to efficiency. Furthermore, their board members become keen advocates of their cause. They constitute extremely effective methods of orienting decisions.

Associations

73. As this a subject dealt with in the following report, it will not be discussed here. It need only be said that Associations are a fundamental anticipatory instrument and that they are bound to play an increasingly important role.

V

PUBLIC PARTICIPATION AND ENVIRONMENTAL MATTERS[1]

SUMMARY

Many environmental controversies involve fundamental disputes over social and political values, over contradictory public attitudes toward growth and conservation, and over different levels of social acceptability to risks and uncertainty. Issues such as these can be dealt with effectively and democratically only through broadened public discussion and debate. Therein lies one of the essential goals and values of public participation.

Government has sought to respond to demands for increased public participation in decision-making on environment-related issues in a number of ways: new consultative measures have been adopted; administrative and regulatory decision-making processes have been reformed; environmental assessment and review procedures instituted; public hearing and inquiry processes expanded; and new ad hoc mechanisms experimented with. The results of these different measures have been generally positive.

National experiences with the promotion of increased public participation have also revealed a number of common problems. The most obvious one is that participation takes time; it is not always conducive to "efficient" decision-making. This suggests the need for more effective mechanisms, ones which will help to reduce lengthy and costly efforts aimed at public consultation, ones which will still permit the fullest possible search for political and social consensus.

Special attention should also be given to improving institutional approaches by which citizens can be more directly involved in the choice of policy goals before debates on specific plans are launched. Efforts to engage the public at the earliest feasible point in the policy planning process provides an opportunity for assessing public desires

1. This report is a survey and an analysis based on experience in siting and developing certain types of facilities in some Member countries.

and needs, clarifying elements of controversy, and evaluating the full range of policy options.

Information is a prerequisite to effective public participation, and governments have a responsibility not only to make information on environmental matters available to the public in a timely and open manner, but also to ensure that citizens are able to provide constructive and timely feedback to government. Under certain circumstances, government financial assistance to citizen groups may be advisable for the purposes of aiding them in developing their own expertise.

One of the central challenges posed by demand and need for increased public participation in environment-related matters is that of integrating scientific and technical facts with political and social values. There are, of course, no easy, simple or proven methods for resolving this problem. Nevertheless, public participation can be seen as an essential means for increasing environmental as well as political awareness, for clarifying the choices to be made, and for seeking social consensus on the balance to be sought between economic development and environmental concerns.

PUBLIC PARTICIPATION AND ENVIRONMENTAL MATTERS

I. INTRODUCTION

1. Demand for public participation in environmental matters is not an isolated phenomenon; it is but one manifestation of the growing desire on the part of many citizens for more immediate and direct involvement in the conduct of public affairs. The public that demands participation in decision-making on environmental matters is comprised of many heterogeneous elements. The objectives sought vary widely and sometimes conflict. The distinction between participation and protest is often obscure. And yet, over the past decade, active public participation has not only served to increase general public awareness of environmental affairs, but has contributed significantly to the elaboration, design and implementation of environmental policies and protection measures. It has also led to considerable controversy, complexity and, in some cases, to increased costs and delays in governmental decision-making.

2. This report seeks to clarify some of the factors behind emergence of new participatory demands, examines how governments have sought to respond to them, identifies some of the problems they have encountered, and suggests a number of measures that might be taken for the promotion of more effective public participation in environment-related decision-making. [1]

A. Organisations and Objectives

3. One hallmark of contemporary industrialised countries is the increasingly large number of special interest groups placing new demands on society to respond to their often particular needs and desires. The rapid growth of national, regional and local environmental groups is

1. This report is based in part upon findings from a more detailed study, carried out under the auspices of the OECD Committee for Scientific and Technological Policy, entitled: "Technology on Trial: Public Participation in Decision-Making Related to Science and Technology," (OECD, Paris, 1979).

but one demonstration of this trend. In the United States, where participation in voluntary associations is widespread, membership in major national environmental organisations increased over the last decade by 330 per cent.[1] In almost all other OECD Member countries, the pattern, if not the actual rate of growth, has been similar: national environmental organisations have grown and proliferated and have developed broad-based constituencies.

4. Equally significant has been the emergence of numerous local and community citizen action initiatives. Many of them have been directly concerned with environmentally-related matters. In Germany, more than two million citizens have taken part in various "Bürgerinitiativen" over the last several years.[2] In France, it has been estimated that nearly 300,000 local and neighbourhood "associations" exist, representing the different community interests of French citizens. In the United Kingdom, membership in local amenities groups and other environmentally-related voluntary organisations has risen dramatically since 1965. From the "Green Wave" in Scandinavia to the "Green Nans" movement in Australia, citizens are increasingly inclined to affiliate themselves with specialised groups representing their particular interests on specific issues of environmental concern.

5. There are at least two explanations for this trend. First, and most obvious, is the fact that people are concerned about the general decay of the environment. They see that some of the environmental "amenities" of life are disappearing; they fear that their children's futures may be threatened; and they are concerned about the unseen, and unforeseen, dangers and risks to human health of living in highly technologically-oriented societies. Thus, they see their own interests at stake in environmental matters.

6. Second, many people to-day perceive traditional channels of representation as having become oligarchical. There is an apparent widespread feeling that political parties no longer adequately reflect contemporary socio-economic conditions, and are ill-adapted to respond to emerging social needs, interests and values. People are concerned about trends toward increased concentration of economic and political

---

1. Based upon surveys undertaken by the U.S. Federal Energy Administration, 1974-75. See also: S. Verba and N. Nie, Participation in America:"Political Democracy and Social Equity,"(Harper and Row, New York, 1972) and R. Mitchell and J.C. Davis III, "The United States Environmental Movement and its Political Context: An Overview", (Resources for the Future, Washington, 1978).

2. The German Bundesverband Bürgerinitiativen Umweltschutz, the largest organisation co-ordinating environmental initiatives, covers more than 900 groups with over 200,000 members. See: Battelle Institut,"Bürgerinitiativen im Bereich von Kernkraftwerken," (Battelle, Frankfurt, 1975).

power; power which is perceived by some as resting too much in the hands of too few. They are therefore seeking new ways of expressing their discontent, of exercising their collective influence, and of imposing their views.

7. As new environmental groups have emerged, they have sought to fill a perceived gap in the more traditional constellation of industrial and professional interest groups which are part of the "established" clientele of government administrators and policy-makers. They have demanded greater access to information in the hands of government and more direct participation in decision-making. They have sought to ensure greater degrees of public accountability. They have tried to influence not only the content and outcome of decisions, but also the processes by which the public interest is defined and the decisions are made.

8. Public participation is not, however, a universal phenomenon. There is, at the same time, a greater degree of "activism" on the part of some groups and individuals, and increasing apathy among others. Neither does this participatory phenomenon touch all areas of public policy: it is fluctuating, sometimes issue-oriented and often apparently random.

9. The objectives sought by such environmental groups are manifold. Many citizen action initiatives have developed spontaneously, coalescing around concrete issues of immediate local or community concern; for example, plans for a new sewage treatment plant or energy-generation facility, a new highway or airport. Often their aims are specific and short-term; to oppose a planned project, propose new alternatives or influence its design or location. Sometimes, however, as in many cases involving urban and regional planning, housing, renewal and rehabilitation, such local initiatives are of a long-term nature, as citizens have set up their own community action programmes and sought to gain and maintain a greater measure of influence over decisions affecting their own lives.

10. Although issues of environmental conservation and local community control remain central to many citizen concerns, one witnesses to-day a clear broadening of environmental interests: from preoccupation with issues of land use to concerns about the social and economic aspects of planning; from concern with the local siting of energy facilities to questioning of the aims of national energy policy; and from questions of how to improve present living conditions to debates about the kind of future society in which citizens want to live.

B. Participatory Strategies

11. Political styles and traditions of government decision-making and policy formulation vary widely between countries. Each nation has its own customs and practices which influence, among other things, the degree of openness of governmental processes, the amount of access the public is granted to government information, and the ways in which consensus is sought on government decisions. Therefore, the strategies employed by persons and groups seeking more direct participation in environment-related decision-making depend on the nature of and opportunities provided by particular political systems.

12. In parliamentary systems, such as Australia, Canada and the United Kingdom, where information is often alleged to be the key to influence, environmental pressure group activity usually takes place through existing, traditional political channels. For example, one important political tactic is to establish contacts with civil servants and Members of Parliament and their advisers in order both to obtain relevant government data and to provide officials with information on specific issues and on citizen views and desires.

13. In the more pluralistic, adversary system of United States politics and administration, additional strategies are often used. One of the most important of these is litigation and the threat of litigation. This approach to participation is the result of a number of factors: recently liberalised rules of legal standing for environmental groups: the availability of environmental impact statements (EIS), the adequacy of which can be tested in court; and the emergence of environmental law firms. [1]

14. In a number of Western European countries, new political parties have emerged specifically oriented towards ecological concerns. Some have been successful especially in recent state and cantonal elections. Equally important, these new ecology parties have caused traditional political parties to consider broad environmental issues more seriously, and to include such concerns in their political platforms. In all countries, other participatory strategies of course have been employed, including the organisation of public protests. A common aim, then, of these different participatory approaches has been to: mobilize

---

1. The emergence of non-profit, tax-exempt public interest law firms specialising in environmental law matters has been facilitated by changes to U.S. Internal Revenue Service tax law (1970) and initial financial support from several U.S. philanthropic foundations. In addition to the two major U.S. environmental law firms, the Environmental Defense Fund (EDF) and the National Resources Defense Council (NRDC), the Sierra Club has established a Legal Defense Fund and the National Wildlife Federation, a legal assistance department.

citizens, politicise issues and influence public opinion, and thus affect the outcome of governmental decisions.

15. As environmental interests have broadened over the last decade, participatory strategies have themselves evolved and matured. Confronted with growing public anxieties over inflation and unemployment, environmental groups in some countries have quickly come to realise that self-righteous and moralistic forms of argumentation are poor substitutes for facts and technical evidence. They have engaged full-time professional staffs, developed their scientific and technical skills, and begun to challenge government and industrial authority and technical expertise. They have also begun to develop strategic coalitions on issues of common concern with professional societies, minority groups and industrial trade unions.

C. Government Response

16. Public participation in government decision-making covers a broad spectrum of activities: from informing and consulting citizens and policy-makers on decisions, to government delegation of decision-making power to citizens themselves. One distinction that needs to be made therefore is between a) processes of public participation in which final decisions are taken by the elected representatives, and b) those in which the outcome is decided directly by the citizens. This report concentrates largely upon the former; that is, on government efforts to promote participation within the context of existing processes of representative decision making. [1]

17. Government has sought to respond to demands for public participation in a number of ways: new consultative measures have been adopted; administrative and regulatory decision-making processes have been reformed: environmental assessment and review procedures instituted; public hearing and inquiry processes expanded; and new ad hoc mechanisms experimented with. Nevertheless, governmental response to environment-related participatory demands has been cautious.

18. One reason for this is that the "environmental movement" itself is imperfectly understood. It is an assembly of many individuals and groups with similar concerns but often different approaches to the "resolution" of environmental problems. Because some environment-related organisations do not fit the landscape of traditional political pressure groups and lobbies, they cause consternation. Because they do not share common sets of ideologies or always abide by traditional

1. Included in this context is the use of direct democratic decision-making procedures, such as public referenda.

political tactics, they are sometimes worrisome. And, because many attract support from increasingly large numbers of citizens, they are often seen as threatening to the "established" political order.

19. For many elected representatives and appointed government officials, there is an additional set of unknowns that cause concern and hesitation. What are the possible consequences of expanding public participation? What impact will they have on decision-making efficiency and on existing administrative and political structures and processes? Are there practical limits to public participation, and what are their implications for the concept of representative democracy?

20. This report reviews some of the more recent government initiatives taken in response to participatory demands. We examine public participation, first, in national environmental decision-making. We analyse various techniques to promote participation in government policy review and formulation and in decision-making on specific environment-related projects and proposals. We then examine participatory techniques used in local and regional planning, focusing on initiatives to involve the public in physical planning matters and municipal decision-making. In the last section of this report, we suggest a number of issues that need to be considered in developing more effective approaches to public participation in environment-related government decision-making.

II. NATIONAL ENVIRONMENT-RELATED DECISION MAKING

21. Some of the more extreme and vocal demands for public participation are being felt at the level of policy formulation and implementation. This is perhaps a reflection of both the inadequacy of existing democratic institutions to respond effectively to new and emerging public preferences and values, and of the lack of social consensus which characterises the fragmented body politique of many contemporary societies.

22. In many areas of public policy debate on environmental matters today, one finds sectors of the general public in opposition. Qualitative concerns for preservation and conservation conflict with the exigencies of growth and consumption; local demands for self-determination are viewed as threatening to the national welfare; and pressures to preserve cultural and economic identity are opposed by the forces of assimilation.

23. Some governments have, by and large, come to accept the inevitability of the fact that only through the promotion of more intensive public discussion and debate can many of these conflictual issues be resolved. They have sought through a variety of means to better inform the public on the implications of the political, economic and value choices inherent in environmentally-related decision-making, to more adequately assess

the full impact of decisions, and to try to resolve conflicts over their implementation. We examine, below, a variety of different participatory techniques to involve the public in national environment-related decision-making.

A. Public Information Programmes

24. Government attempts to improve levels of public understanding on complex environment-related matters have taken a number of different forms. Some have been highly centralised with government assuming primary responsibility for their design and implementation. Others have been more decentralised relying largely upon the efforts of local groups and non-governmental intermediary institutions.

25. Several of the most recent examples are to be found in the area of energy policy and planning, and specifically in relation to civilian nuclear energy programmes. The example of nuclear energy is especially insightful for two reasons. It reveals the types of difficulties governments often encounter in attempting to inform the public in areas of extreme technical complexity and political controversy. It also illustrates the dilemma of integrating scientific and technical facts and expertise with political and social values.

26. In 1974, Sweden initiated a one-year programme of national discussion and debate on energy matters, involving the use of study circles. Denmark and Austria, in 1974-75 and 1976-77 respectively, also undertook national energy information campaigns. And Germany, beginning in 1975, started its "Dialogue with the Citizen" (Bürgerdialog) on national energy policy and planning.

27. Each of these initiatives was undertaken on an ad hoc basis and for a variety of reasons, although all were in response to the same general "problem"; growing public controversy surrounding civilian nuclear energy programmes. All were aimed principally at promoting more informed public debate by improving levels of technical understanding, although each took a somewhat different approach to this task.

28. For example, the Danish-sponsored non-governmental Committee on Energy Information concentrated heavily in its campaign on encouraging "grass roots" initiatives through the distribution of grants to a large number of individual regional and local groups. Some of these groups focused primarily on informing themselves on alternative energy sources and not on nuclear energy matters. By contrast, the Austrian government-administered energy information campaign focused principally upon nuclear-related issues, through the conduct of a series of nationally-televised debates between nuclear proponent and opponent experts.

29. In all of these countries but Germany, the energy public information programmes were terminated after one or one and one-half years of operation. Only in Austria, where a national referendum on nuclear energy was subsequently held, was there an attempt to link directly information campaigns to a national vote.

30. More recently, Canada and France have also begun to experiment with the use of energy information committees comprised of members representing divergent views on nuclear energy. While, in the Netherlands, the government has initiated plans for a so-called "broad societal discussion" on nuclear-related issues in an attempt to generate a general social and political consensus on future national energy policies. In New Zealand, following a recent Royal Commission inquiry concerned with the introduction of nuclear power, the government has circulated a report on "Energy Goals and Guide-lines" in an effort to encourage wider public discussion and debate on national energy policy matters.

31. It is difficult to assess the value of these different programmes in terms of their impact on improved environmental or technical understanding. Public opinion poll findings show somewhat contradictory results. While these polls may accurately portray shifts in public attitudes towards, for example, nuclear energy, they hardly suffice as an indicator of public understanding or technical appreciation. What these information programmes did undeniably achieve, was increased public awareness of the broad set of interdependent factors - environmental, social, economic and technological - surrounding current national energy debates. They also revealed some of the difficulties inherent in government attempts to educate or inform the public on controversial matters where public opinion is already highly polarised.

32. For example, one of the major difficulties encountered by government institutions furnishing information in these campaigns was that, because their own institutional credibility was often questioned, so too was the information they provided. Because information is often believed to be only as credible as its source, a major barrier to government is to overcome public scepticism about the objectivity of the information presented and the motivations of government as sponsor.

33. Public information campaigns also provided an ideal target for those citizens and activist groups seeking to give broader public currency to their own particular views. Experiences with government-initiated energy information campaigns and committees provide a number of examples of this, and also attest to the difficulty of trying to narrow prematurely the scope of discussion, without fully grasping the roots of controversy.

34. Efforts to promote more informed public debate, whether in the field of nuclear energy or in other areas of considerable public controversy, require that the strategies adopted fit the objectives sought. If

the aim of an information campaign is to defuse political controversy, it is unlikely that technical facts and forms of argumentation are the most appropriate tools for the task. Similarly efforts to achieve social consensus are severely impeded if there is no consensus in the first place on what is in disagreement.

35. These problems suggest the importance and value of efforts aimed at encouraging pluralistic and decentralised approaches to citizen self-education and public opinion formation. Timing also plays a critical role in most efforts aimed at promoting broader public understanding. As these experiences suggest, trying to implement a government public information campaign once public opinion is already polarised and an issue highly politicised presents an almost insurmountable task. In some areas, such as those associated with energy policy, this has been difficult to avoid, owing in part to the suddenness with which energy matters have come to dominate national concerns. Nevertheless, efforts to promote early public discussion and understanding do not amount merely to practising political preventive medicine, but to trying to ensure that future debates will provide the necessary basis for broadly acceptable decisions.

36. Finally, and most importantly, one must place the problem of public information and understanding (or the alleged lack thereof) in its proper perspective. For it would be too simplistic to conclude that opposition or protest in areas such as nuclear energy can be attributed solely or even mainly to insufficient public knowledge and understanding. Many of those who oppose certain technological programmes are extremely well-informed as to their details and associated risks. Nor can attitudes of public scepticism be explained by one reason alone. Often they result from a broad panoply of public concerns and misgivings about the social goals to be pursued, the environmental protection measures to be taken, and the way costs and benefits are to be distributed across society-at-large. Moreover, access to information without effective means for its use is like possessing a lever without a fulcrum. As has often been said: information is a necessary but insufficient condition for public participation.

B. Commission of Inquiry

37. Conflicts between local and national interests lie at the heart of many environment-related public controversies. The siting of industrial or energy-related facilities, the building of airports, highways and dams, - these are some of the more obvious and controversial examples of projects that impinge on local interests but yet may serve the general public welfare. In many cases their impacts can be adequately assessed and the conflicts they raise resolved satisfactorily

through traditional parliamentary and administrative analysis and review procedures, especially those that require the preparation of environmental impact statements. In others they cannot.

38. Many extremely large-scale technological undertakings- such as in the area of uranium mining, spent nuclear fuel reprocessing and storage, natural gas pipeline construction and electric power planning- have extensive and multiple social, environmental, economic and political impacts, the full assessment of which poses new and difficult challenges. Such multi-dimensional impact assessments require considerably study, expertise, consultation and, most of all, time.

39. Parliamentary bodies are by necessity preoccupied with problems of immeditae political concern. Legislators often lack the time, staff and resources required for sustained and detailed enquiry into such complex matters. Moreover, political opinion on many such issues is often polarised, producing cleavages running through most major political parties.

40. In a number of countries, therefore, governments have taken recourse to another, ad hoc mechanism - commissions of inquiry - to assist in the study and assessment of large-scale technological projects and complex public policy issues. Such commissions can serve as one means both to secure needed technical advice, and to create a climate of opinion within which parliaments can exercise their prerogative to decide.

41. Commissions of inquiry are not new; they are one of the oldest institutions of public consultation in parliamentary systems of government. They have a long tradition of providing considered advice to government on a broad range of issues. In recent years, however, they have taken on new and broader functions: that of opening up oftentimes controversial issues to widespread public discussion; of providing more representative opinion to decision-makers on the different needs and wishes of the people; and of assessing the impact of various policy alternatives before final policy choices are made.

42. Commissions of inquiry can be seen, therefore, as complementing existing means for assessing the full range of potential impacts of a proposed project and/or of public opinion on a given issue. Moreover, some issues more than others directly affect minority group interest - such as those of Aboriginals in the case of Australian uranium mining or of Inuit (Eskimos) and Indians in the case of northern gas pipeline construction in Canada - which have heretofore been under-represented in most government policy making forums. The Commission mechanism is thus also often considered not just as a mechanism for general public participation, but also as an important means for promoting minority

group involvement, and for trying to reconcile these and other competing group interests with those of the general public.

43. The so-called Berger Inquiry, established to examine a proposed 5,000 km. natural gas pipeline project through the Mackenzie Valley from the Arctic Circle to the southern Canadian/U. S border, contained a number of innovative participatory procedures for the comprehensive assessment of environmental impacts.[1] Four different sets of public hearings were organised. Formal hearings were established to receive testimony on the pipeline construction proposals and to examine engineering problems as well as evidence as to project impacts on the physical, living and human environment. Community hearings were organised and were held concurrently with the formal hearings to allow residents along the proposed pipeline corridor to express their views on the project. A number of special hearings were also organised to investigate specific problems relating to northern gas exploration and production activities. And, finally, a series of southern hearings were held in metropolitan centres across southern Canada to allow Canadians who could not appear at the northern hearings to express their views.

44. The community hearings, in particular, were a major departure from previous public inquiry procedures in that they were largely informal, and permitted average citizens and native people to express their concerns openly, freely without having to submit to formal cross-examination. In addition, summaries of evidence presented at the formal hearings were broadcast by radio in six native languages to northern communities, thus allowing citizens to stay abreast of and respond to these "formal" findings in the community hearings.

45. A second major innovation of the Berger Inquiry was its financial assistance to intervenor groups. The Canadian government provided nearly $ 500,000 to the Inquiry for support of research and studies carried out by environmental and other interest groups. An additional $ 1 million of government funds was also given directly to native groups to permit them to undertake research relating to outstanding land claims and to prepare testimony for presentation at the community hearings.

46. A third innovation was to require that the Commission Counsel serve as an active, independent participant in the hearings with powers to obtain any evidence he deemed relevant to the Inquiry. He was also directed to present his arguments and views publicly; the aim being to discourage the impression that the Commission Counsel or Inquiry staff were serving as a kind of "privy council" to the Inquiry.

---

1. See: T.R. Berger, "Northern Frontier-Northern Homeland: Report of the Mackenzie Valley Popeline Inquiry," Vol. 1 and 2, (Supply and Services Canada, Ottawa, 1977) and OECD, "Technology on Trial," op. cit.

47. Finally, and perhaps most importantly, the Inquiry took its time. The Inquiry began in the late spring of 1974 with a series of preliminary hearings and ended in the fall of 1976. The final Inquiry report was completed a year later in November 1977.

48. The innovations introduced in the Berger Inquiry have, to a large extent, been incorporated in other Canadian public inquiry procedures. The approach to funding of intervenor groups has also received careful consideration in a number of other countries. One of the merits of combining formal with informal community hearings was the amount and diversity of information and views available to the Inquiry. It also permitted, and in many cases demanded, a careful integration of technical, empirical facts, with subjective social, political and cultural values; a process of integration often lacking in more "traditional" public inquiries into the potential impacts of large-scale technological undertakings.

49. The Berger Inquiry, along with other inquiries such as the one conducted recently in the United Kingdom concerning spent nuclear fuel reprocessing at Windscale, also demonstrate the value of adversarial approaches to the examination and cross-examination of evidence. In both cases, this led not only to more critical information becoming available, but to the more thorough and penetrating examination of competing claims and to the clearer articulation of individual and group interests and biases.

50. Commissions of inquiry can be, nevertheless, quite costly and time-consuming.[1] Their effectiveness as a participatory mechanism is often heavily dependent upon the qualities of leadership and direction provided by the person or persons appointed to conduct them. Moreover, because commissions of inquiry are in most cases advisory, their recommendations are subject to final ministerial and cabinet approval. Thus, public participation in inquiry proceedings, while potentially influential, is only indirectly, and not directly, related to the actual exercise of government decision-making power.

51. Nonetheless, within parliamentary systems of government commissions of inquiry can complement in a number of important ways the traditional means for examining the implications of medium and long-term policies, for anticipating potential adverse environmental impacts of policy intentions, for marshalling, the resources of time, objectivity and expertise required for coping with issues of extreme technological complexity and diversity, and for providing an open forum for the expression of public opinion.

---

1. For example, the Canadian Mackenzie Valley Pipeline Inquiry lasted nearly 3 years and is estimated to have cost about $ 5 million, while the Australian Ranger Uranium Environmental Inquiry took about 1 1/2 years and cost nearly $ 1 million.

C. Environmental Assessment Procedures

52. Environmental impact and assessment review procedures represent one of the most important developments in relation to public participation in many countries over the last decade.[1] The preparation of environmental impact statements, which originated with the U. S. National Environmental Policy Act in 1969, are now required or under consideration in many OECD Member countries.[1]

53. Each year in the United States more than 30,000 proposed federal actions are reviewed to determine whether or not the agency concerned will be required to prepare an environmental impact statement (EIS). Since 1969, nearly 8,000 EISs have been filed with the Council on Environmental Quality.

54. Impact statements serve two important purposes: they inform decision-makers of the likely effects of proposed actions and they provide the public with the opportunity to express their views concerning those actions. They have also opened the door to some litigation. Between 1970 and 1975 there were 332 suits filed against Federal agencies for inadequate EIS reports, of which 64 were sustained. Despite the fact that some 95 per cent of draft EISs escaped legal challenge during this period, the threat of litigation remains a powerful inducement favouring participation.[2] For many people the contribution of environmental impact statements to more informed public debate outweighs their disadvantage as a tool for litigation.

55. In Canada, the Federal as well as most provincial governments have also adopted environmental impact statement procedures, and they have also served as a means to broaden mechanisms for public consultation. The Canadian government has established a number of special environmental assessment panels to hold public hearings on issues ranging from nuclear power plant siting to liquid natural gas (LNG) transport and spent nuclear fuel refining and waste management facilities.[3]

56. Similarly in Australia, the use of environmental impact statements has served as an important instrument for involving the public in environment-related decision-making.[4] France has also recently adopted

---

1. See: OECD, "Environmental Impact Assessment," (OECD, Paris, 1979).

2. Council on Environmental Quality, "Environmental Impact Statements: An Analysis of Six Years' Experience by 70 Federal Agencies." (GPO, Washington, March 1976).

3. See, for example: "Environmental Assessment Panel, Report on the Port Granby Eldorado Nuclear Ltd.", Uranium Refining and Waste Management Facility. (Department of Fisheries and the Environment, Ottawa, 1978).

4. See, for example: "Report of the Ranger Uranium Environmental Inquiry" (Fox Report), "Final Report," (AGPS, Canberra, 1977).

procedures for the preparation and review of such statements, and this should provide increased opportunities for citizen participation in decision-making.

57. In each of these countries, environmental assessment and review boards and panels do not have the power to prohibit government authorities from implementing any proposed action, no matter how grave the environmental consequences. These decisions are ultimately taken by the chief executive or cabinet and, in some countries, by a vote of Parliament.

D. Regulatory Decision-making

58. Historically, adjudicatory procedures required for the granting of licences and construction certificates and for the setting of regulations and standards governing such activities as airport and pipeline construction, power plant siting and off-shore oil drilling involved only a limited number of parties. Increasingly over the last decade, as new minority groups have emerged, mobilising sometimes around single issues, they have achieved access to regulatory boards, multiplying considerably the number of complainant parties involved in such proceedings.

59. In many countries, there has also been a trend towards requiring all regulatory hearings to be held in public. However, each country responds differently to the question of which person or groups may participate in these hearings and what issues may be contested. In Denmark and Germany, the issue of "standing" - the right to participate as "full parties" in regulatory proceedings - is usually limited to those parties having an "individual, substantial interest". This is often defined in financial or material terms, or in terms of the person's or group's geographical proximity to the site of the proposed facility or project under regulation. This has meant that for those groups not meeting criteria for standing, the only way to become indirectly involved in regulatory disputes has been by providing assistance to those persons having a right to intervene.

60. In other countries, such as Canada, regulatory boards have exercised broad discretion in granting standing. The result has been that, generally speaking, standing has not so far posed a major barrier to citizen or group participation in Canadian regulatory proceedings. In the United States, however, some of the major battles waged before regulatory bodies have been fought over the very issue of standing. Most U.S. regulatory commissions and boards retain considerable discretionary powers to limit intervenor presentations.

61. One direct result of liberalising rules of standard and encouraging a broader representation of public interests is that more groups and individuals than ever before are now becoming actively involved in regulatory proceedings. This in turn has led to some concern on the part of regulatory bodies as to whom various intervenor groups represent. This concern has become especially acute with respect to "third party" environmental and so-called "public interest" groups, who may represent some affected interests but who are not themselves directly or sometimes even indirectly affected.

62. Some people believe that "representativeness" is somewhat of a misnomer with respect to regulatory matters. Whom one represents is, of course, important when an individual or group purports to speak as a proxy or surrogate for others. Knowledge of one's constituency is especially appropriate with respect to the exercise of political power in an electoral system where persons are chosen to represent the views of their fellow citizens. However, regulatory bodies are not, in principle, political in this sense; theirs is not an electoral function. Their role ought to be to consider all views, irrespective of whether they are shared by few or by many, to decide whether each point of view has merit, and if so, to take it into account.

63. Regulatory bodies, such as energy boards and commissions, have also become increasingly engaged in the difficult and hazardous task of long-range forecasting, so as to be able to establish appropriate rationales for policy recommendations. This has meant that while regulatory bodies continue to rely primarily on proponent and opposition parties to produce and develop factual materials, they are no longer confined solely to the role of evaluating these facts. They are assuming increasing responsibilities for independent fact-finding.

64. The significance of this for public participation in regulatory proceedings is that intervenor groups must establish the necessary expertise to participate effectively in regulatory proceedings. This can be both time-consuming and costly. Many citizen groups find themselves at a disadvantage when confronting industrial proponents who who can afford to hire the best available technical expertise whose costs may be either tax-deductable or can be recovery by product price increases.

65. This has raised an important and controversial issue as to whether or not public funds should be provided to assist citizen groups in their interventions before regulatory boards and tribunals.

---

1. See: G. Goetz and G. Brady, "Environmental Policy Formation and the Tax Treatment of Citizen Interest Groups", in Law and Contemporary Problems. Vol. 39, No. 4, (USA), 1975, pp. 211-231.

66. Both sides in this debate seem to agree that more knowledgeable public participation in government decision-making is desirable. However, some fear that direct government financial assistance to citizens' groups for the purposes of acquiring information and technical expertise will only lead to further citizen intervention and eventually to the disruption of governmental decision-making processes; that it will not only precipitate deadlocks, increase workloads of government agencies and courts of appeal, but add a new dimension of opposition and frustration. In sum, these people believe the public interest to be better served by government itself; by its elected and appointed officials through the exercise of their legislated and assigned responsibilities.[1]

67. Others do not necessarily disagree with this interpretation. They argue, however, that government must always seek to ensure that identiable special interests shall be heard; that while most citizens' groups cannot in fact represent the interests of the whole general public, it is in the public interest that their views and interests be put forward. Direct financial assistance to such groups is believed to be especially important in those policy areas involving complex technological issues, where individual and group interests are not often readily identifiable or fully represented.

68. The costs associated with citizen participation in regulatory proceedings can vary widely depending upon its nature, its duration and the complexity of the issue in question. In the United States, for example, it has been estimated that the cost of active participation in rule-making at the Food and Drug Administration (FDA) is in the range of $30,000 to $40,000. At the U.S. Interstate Commerce Commission (ICC), estimates have ranged as high as $100,000 for full-time participation in licensing proceedings.

69. A number of U.S. regulatory agencies have sought to lower this barrier to public participation by developing compensatory programmes of financial assistance to citizen groups. These include, among others, the FDA, the Federal Energy Administration (FEA), and the National Highway Traffic Safety Administration of the Department of Transportation.[2] In addition, the U.S. Toxic Substances Control Act of 1976 also authorises the Administrator of the Environmental Protection Agency (EPA) to provide compensation for expert witness fees and for the costs of participating in rule-making proceedings.[3]

---

1. For representative views on each side of this debate in the United States, see: U.S. Senate, Committee on Government Operations, "Public Participation in Government Proceedings Act of 1976," Hearings. (GPO, Washington, 1976).

2. Others include the Civil Aeronautics Board (CAB) and the Consumer Product Safety Commission.

3. "Toxic Substances Control Act of 1976," 15 U.S.C., para. 2619.

70. One EPA spokesman has noted the contributions that citizen representatives have made to agency decision-making: "They have provided facts and arguments relevant to the statutory purpose (of EPA) that would not otherwise have been urged on the agency and which the agency might not have uncovered or fully appreciated on its own."[1] Nevertheless, there still remains considerable ambivalence about whether or not government should provide direct financial assistance to citizen groups engaged in regulatory proceedings.

E. Administrative and Judicial Appeal

71. Citizen recourse to the legal system as a means for resolving environmentally-related conflicts has become prevalent in some countries. Citizen groups have sought to block the construction of energy facilities, sue private industries for developing technologies in violation of environmental standards and contest governmental enforcement of environmental regulations. Legal disputes have also developed over the issue of weather modification, at the same time that questions have been raised about the legal responsibilities of researchers working on earthquake prediction. Biomedical research and innovation has been the subject of several major legal disputes involving concerns about bio-hazards.

72. The use of legal channels as a form of direct public participation in decision-making is motivated by a number of factors. Some citizen groups consider legal recourse the only remaining alternative to overt public protest. Dissatisfied with their influence through public regulatory channels, they have appealed to the courts to overrule government decisions and judgments. Others have sought legal recourse as a means of clarifying constitutional ambiguities surrounding questions of personal freedom and choice. Still others have sought to redress perceived inequities in the sharing of certain environmental risks by seeking to establish legal precedents in areas heretofore not covered by existing legislation.

73. In many countries, administrative courts and appeal boards have become increasingly involved in judging not only the integrity of administrative processes - their fairness and adherence to statutory responsibilities - but in reviewing the adequacy of evidence to support administrative decisions as well as the merits of those decisions.

74. As in the area of regulation, rules of standing have been modified and liberalised in many countries. This has permitted increasing numbers of individuals and citizen groups to present grievances to administrative courts and appeal boards. In addition, environmental litigation

1. Quoted in "Beyond Sunshine", Trial Magazine, June 1977, p.44.

has brought pressure to include, as cause for standing, not only damage to property rights, but some of the more "intangible" health and aesthetic impacts of technological progress. How accessible is the legal process to the average citizen, and what opportunity does it provide for him to express his concerns?

75. The costs of litigation can be inhibiting.[1] This is especially true when the risks are high that, if unsuccessful in their appeal, the citizen or group will have to bear all legal costs. Moreover, the private foundation support to public interest law firms found in the United States is generally not available in most other countries.[2] Groups must depend upon private donations, which are often insufficient to cover the costs of litigation and those associated with research and hiring of experts.

76. Also, legal involvement often means that case preparation and its defence are predicated upon legal assessments of the most efficacious tactics to employ, and not on public assessments of what is "important" to debate. And as one U.S. lawyer has pointed out, oftentimes, "the intervenor becomes the captive of the lawyer".

77. Nevertheless, citizen recourse to legal channels represents a significant development in the whole process of public participation. It reflects a rendency in some countries to try to reconcile important political, socio-cultural and ethical conflicts by recasting them in a juridical frame of reference. It is also, perhaps, one indicator of some of the shortcomings of existing forms and mechanisms for public participation. For many would largely agree that recourse to litigation as a means for settling environmental disputes is not only time-consuming and costly, but does not always lead to satisfactory results for all parties concerned. Increasing recourse to environmental litigation should therefore serve as a warning about the need to develop more timely, open and effective processes of public participation.

F. Other ad hoc Participatory Mechanisms

78. Over the last decade, a certain number of ad hoc participatory mechanisms have also been devised and experimented with to cope with and promote public involvement in nationally-related environmental disputes. Environmental mediation and arbitration techniques have

1. In some countries, attorney's fees are awarded to the victor, while in others, the unsuccessful party must bear all legal costs. For further details on the rather special conditions prevailing in the United States, see: Council for Public Interest Law, Balancing the Scales of Justice: Financing Public Interest Law in America, (Washington, 1976).

2. For example, during the early 1970s, the Ford, Rockefeller and Clark foundations provided "seed money" to a number of U.S. public interest law firms on the understanding that those firms would soon be financially self-supporting.

been used in a number of countries to resolve specific conflicts at the interface between national environmental policies and their implementation at local levels. Local consultative referenda - and in several countries, state and national referenda - have been held on a variety of issues of environmental concern. And, a large number of techniques have been employed to more accurately assess public opinion with respect to proposed environmental protection measures.

79. One technique developed and experimented with by the U.S. Forest Service since 1972 is a system called "Codinvolve", an analytical tool permitting policy-makers to evaluate a broader spectrum of attitudes and public preferences regarding environmental protection matters than may emerge at public hearings or in response to public announcements. This is an applied content analysis system designed to transfer large quantities of information from written statements in diverse sources into a form that is easily summarised for policy-makers.

80. After identifying questions of significance to decision-makers, researchers code all relevant public statements appearing on a given issue in, for example, personal letters, reports, petitions and newspaper editorials. Condensing these various statements into clusters of opinions, tables are then prepared for policy-makers which summarise the nature of public inputs in terms of degree of support or opposition expressed for a given issue or proposed project.

81. The Forest Service's use of "Codinvolve" data on public attitudes about the use of DDT provides one example of the application and limits of this technique.[1] In its environmental impact statement prepared for the U.S. Council for Environmental Quality (CEQ), the Service requested authorisation to use DDT to protect Douglas-Firs in forests in north-western United States against moth damage. It provided "Codinvolve" data indicating considerable public support for DDT use, owing to the general unavailability of other effective alternatives and concern about the economic implications of moth infestation. However, the CEQ and Environmental Protection Agency (EPA) ruled against using DDT, contrary to the wishes of local citizens, giving greater weight to the concerns of environmentalists, who were strongly opposed to DDT use.[2]

82. Such applied social science research techniques to measure public sentiment are increasingly being used in a number of countries to

---

1. See: Kelley and Rompa, "Public Opinions about Controlling the 1973 Douglas-Fir Tussock Moth Outbreak", (mimeo), (Portland, Oregon, U.S. Forest Service, Pacific Northwest Region, 1973).

2. A year later, following a second moth outbreak EPA authorised a one-time use of DDT.

complement other more traditional means for informing decision-makers of citizen preferences. They are participatory, however, only in the narrowest and most indirect sense; letters, petitions and questionnaires are only solicited statements of beliefs. They do not necessarily represent a commitment to citizen participation in decision-making.

83. Another set of techniques to gauge public preferences and opinion on national environmentally-related policy issues involves the use of various media-based balloting devices and, of course, public referenda. Radio, television and press opinion poll surveys provide on indication of public attitudes to guide policy-makers in their deliberations on given issues. For many members of the general public, however, such techniques, are, again, considered to be participatory only in the narrowest sense. Increasing recourse to referenda can therefore be seen as one response to the public demand for more direct forms of consultation and direct participation.

84. Public referenda are generally of two types: constitutional which are binding on decisions and consultative which are discretionary. Consultative referenda have been only rarely used in most countries, but have become the subject of increasing attention in a few. As mentioned previously, national referenda have been held in Austria in connection with energy policy and, of course, are a central part of Switzerland's traditional democratic process. In the United States there has been a resurgence of use of state referenda. While, for exemple, in Belgium and France, consultative referenda have been increasingly employed at local government levels to gauge public opinion on a variety of local environmental issues, ranging from the siting of autoroutes and energy facilities to new industrial complexes.

85. Nevertheless, many people have come to realise that public referenda, like most democratic methods, are imperfect. They can be very costly and time-consuming, and are often subject to manipulation. In areas of considerable technical complexity, referendum issues are often grossly oversimplified and are portrayed in such ways as to leave little room for political negotiation or compromise; they are inevitably reduced to questions of "either" and "or".

III. URBAN AND REGIONAL PLANNING

86. Issues involving physical planning have been at the centre of debate for well over a decade in most OECD Member countries. No other issue concerning public participation has received perhaps as much widespread public attention, governmental reaction and academic

analysis than that of how to involve effectively the general public in decision-making on urban and regional planning.[1]

A. Local Land-use Planning

87. Since the mid-1960s, in response in part to aggravated inner city problems and to the effects of a previous decade of urban deterioration and suburban growth, major reforms have been undertaken in many countries to make local physical planning processes more responsive to citizen inputs. In several countries, such as Denmark, Germany and the United Kingdom, participatory procedures have been introduced to encourage early citizen involvement in the drafting of local land-use plans.

88. In the United Kingdom, for example, fairly detailed procedures now exist by which citizen views are sought prior to the drafting of local land-use plans. Once the draft plan is published and if there are valid outstanding objections, the district planning authority can hold a public local inquiry.

89. Similarly, in Germany, in the elaboration of community land-use plans and zoning ordinances, preliminary public hearings are held and the goals and possible effects of plans are discussed before detailed plans or ordinances are drafted. In addition, in the case of zoning ordinances, a "social plan" is required to assess possible social impacts of an ordinance (for example, on living conditions and employment) and to propose alternative measures for their mitigation.

90. In most other countries, participatory procedures allow for citizen inputs only after draft local plans or ordinances are already drawn up and published. These procedures usually take the form of public hearings or inquiries in which citizens can comment on the proposed plans and state any objections. Usually, however, decisions on approval and implementation of the local land-use plans or zoning ordinances are taken by the appropriate planning authority or other governmental body.

91. The use of "planning cells" or citizen panels is one technique that has been widely experimented with, especially in Germany, for the promotion of broader public involvement in municipal government decision-making.[2] This consists of instituting ad hoc panels of citizens,

---

1. See, for example: Council of Europe, "Urban Environment and Participation" (Council of Europe, Strasbourg, 1978) and studies carried out under the auspices of the Council's Steering Committee for Regional and Municipal Matters.

2. See: P.C. Dienel "Pour les Nouvelles Structures de Participation", in Futuribles, No. 7, Summer 1976, pp. 279-293.

randomly selected from the local community, who meet together to study and discuss specific municipal plans or projects for a fixed period, varying in duration from three days to three weeks. Citizens are remunerated for time spent in these planning activities and, although they are not delegated full decision-making powers, they work closely with municipal planning authorities in the preparation of local planning decisions.

92. Similarly, in Canada, a number of somewhat less-structured approaches to citizen involvement in planning-related matters have also been employed. For example, in Ontario, the publicly-owned provincial utility company, Ontario Hydro, has established a number of "citizen committees" and "working groups" comprised of participants drawn from community agencies and interest groups, to assist the utility in the selection of future power plant sites and transmission corridors. These activities have evolved over time from a purely "informational" approach to more active efforts at involving local citizens in "consultation" and "joint planning" study processes.

93. Finally, a more recent technique which has been developed in the United States relates to "citizen review boards". These are bodies whereby decision-making authority is delegated to citizen representatives who are either elected or appointed to sit on a board with authority to review alternative plans and decide which plan should be implemented.

B. Regional Physical Planning

94. Participatory procedures in regional planning, by contrast, are in general more limited. Regional land-use or structure plans, in those few countries such as the United Kingdom where they exist, establish the general framework for the subsequent elaboration of local land-use plans. Thus, while they are not normally binding on citizens, they can limit or pre-empt to some degree future local planning options. Because citizens are generally more immediately concerned with local and not regional planning issues, demand for public participation in this area has been less pronounced. County governments in many countries have therefore taken initiatives to try to encourage participation in regional planning, with often mixed results.

95. The Danish experience with regional planning in Bornholm, a 600 km^2 island with 47,000 inhabitants situated in the Baltic Sea, provides several revealing insights.[1] In late 1975, the County Council

1. See: P.E. Mouritsen, "Public Involvement in Denmark," (Institute of Political Science, University of Aarhus and the Danish Research Administration, Aarhus and Copenhagen, 1977), pp. 66-72.

initiated a six-month campaign aimed at eliciting public reactions on the medium-term (to 1990) objectives of regional planning. The five major elements of this campaign included: an introductory information campaign in which alternative plans were presented in the mass media; mass media presentations focusing on six principal themes ranging from demographic developments to recreation and tourism; a series of public meetings organised by each of the island's five municipalities to debate alternative plans; a series of study circles to further stimulate public discussion; and, finally, in the last three months, a public competition in which individual citizens were invited to submit new alternative regional planning proposals.

96. The intention of the County Council, in undertaking this campaign, was that debate should concentrate more on general, long-term objectives of regional physical planning and not on the details of specific plans. However, in this respect, the debate did not fulfil these expectations. Citizens preferred to concentrate on issues of more immediate and specific concern; for example, bus service, food stores, vacation dwellings and near-term economic and employment prospects. Nevertheless, the campaign did serve to generate a general public consensus on a compromise future urban development plan which was finally adopted by the County Council in mid-1977.

97. Subsequent evaluations of this participatory experiment[1] reveal a number of relevant findings. Citizen knowledge about the county and its problems greatly improved as a result of the campaign. County officials and politicians also learned, through their increased contact with the citizens about citizen needs and the need to express themselves in clearer, simpler language. The experiment served to consolidate and extend the county communication network on planning matters.

98. The experiment also revealed a number of clear citizen preferences as to different channels of participation. Although over a third of the population indicated that they had seen or read the county and municipal planning proposals, discussed them with their fellow workers, family and friends (45 per cent), and listened to the theme debates on the radio (45 per cent), only 4 per cent of the population attended public meetings. And, only 2 to 3 per cent actually participated in study circles or submitted a written alternative proposal to the county authorities. By and large, most citizens preferred staying abreast of planning developments by listening to radio newscasts (70 per cent), and by following newspaper coverage of the campaign (67 per cent).

---

1. Tonboe, J.C. et al., Regionplandebat,"en sociologisk undersøgelse på Bornholm" 1975-76 (SBI, Byplanlaegning 33, 1977).

99. These results also confirm a more general finding of many studies on public participation in physical planning matters. That is, the experiment favoured a certain polarisation of the population: those persons with considerable resources (knowledge, experience and contacts) at the outset of the experiment were most intensively involved in the ensuing debate and, as a result, further enhanced their political resource base. Nonetheless, the experiment did, at the same time, also serve undeniably to broaden significantly general public knowledge and experience on planning matters in Bornholm.

C. Municipal Decision-making

100. One of the most contentious issues concerning participation in planning relates of course to citizen demand for more direct control over municipal decision-making. Although examples of such delegation of government power to citizens are rare, some steps have been taken to try to ensure "maximum feasible" public influence on local planning decisions.

101. For example, as referred to above, local consultative referanda have been organised in a number of countries to gauge public reactions to proposed planning schemes. In addition, attempts have been made to promote the two-way exchange of information between planners and the citizens, by means of panel debates, expanded local hearings and processes of informal citizen consultation. Moreover, as citizen groups have begun to gain access to government planning models and data and to counter-expertise, they have in some instances been able to challenge effectively the merits of proposed government planning decisions and suggest viable alternatives.

102. In most cases, however, final decision authority rests with appointed and elected municipal officials. [1] Despite outspoken public demand for greater degrees of local citizen control, a prevalent tendency in most countries has been to preserve existing processes of representative decision -making; processes whereby government officials retain the ultimate power to decide in what they perceive to be the interest of the majority.

---

1. Two major exceptions include planning experiences in Italy at Bologna and Venice and in Switzerland, where citizens generally retain considerable power over many local environmental planning decisions.

IV. PARTICIPATION AND ANTICIPATORY POLICIES

103. Government has generally reacted to demands for public participation, not anticipated them. In many countries, the response has been limited and cautious. Uncertainty within society as to the different aims of public participation is reflected in government ambivalence as to the degree to which participation should be encouraged, in what fields and under what circumstances.

104. A number of issues are raised which are of specific concern to Environment Ministers. These relate to : the goals that should be devised; how certain problems can be overcome; and what new types of challenges will confront Environment Ministers in the future as they seek to encourage more effective participation in decision-making on environment-related matters.

A. Goals of Public Participation

105. Environmental controversies involve fundamental disputes over social and political values, over contradictory public attitudes toward growth and conservation and over different levels of social acceptability to risks and uncertainty. Issues such as these can be dealt with effectively and democratically only through broadened public debate and discussion. Therein lies the essential aim and value of public participation.

106. Public participation can serve as an opportunity for allowing citizens and groups to express their views and interests, as a political tool for anticipating shifts in public attitudes and values and as a means for building consensus in areas of environmental controversy and conflict. It is not just a form of anticipatory policy but a means for improving the quality and acceptability of government decisions. This is especially evident in the area of environmental impact assessment. In many instances, citizens themselves have considerable knowledge with respect to local, social, economic and environmental conditions which can contribute significantly to governmental processes of assessmen and review. Citizens who are asked to accept certain decisions are often more willing to do so if they have been directly involved in their preparation and given responsibility for their implementation.

107. Public participation can also assist Environment Ministers in clarifying the nature of environmental conflicts between both local and national interests and between competing national policy objectives. While participation may not, in itself, serve to resolve these inherent conflicts, it can serve to clarify the choices to be made and the costs to be borne by alternative policy decisions.

B. New Challenges

108. It is becoming increasingly recognised that public participation, if it is to be productive and successful, has to be a process of two-way communication. It is the responsibility of government not only to make information on environmental matters available to its citizens in a timely and open manner, but also to ensure that citizens are able to provide constructive and timely feedback to government.

109. Because of variations in national traditions and political styles, public participation raises different problems and demands different solutions. Each country will respond as it sees fit. Nonetheless, as has been pointed out in this report, there are a number of common problems that have been encountered.

110. One key problem relates to the general lack of effective participatory mechanisms for involving the public in the formulation and review of environmental policies, as distinct from specific decisions on environment-related projects. With respect to these latter decisions, many new and existing mechanisms for participation in, for example, assessing resource development projects are promising and should be extended. However, public participation in policy review processes remains limited and undeveloped.

111. Special attention should be given to improving institutional mechanisms by which citizens can be more directly involved in the choice of policy goals before debates on specific plans are launched. Such policy review and formulation processes should be carefully structured so as to allow for different stages of public consultation, for data gathering, research and assessment of policy alternatives. All of this implies considerable time, and the need to inform continuously the public on the structure and key decision points in the policy planning process and on opportunities for public inputs and planning information needs.

112. Governments have sought by a variety of means to inform the general public on complex, environmental-related policy matters, often with limited success. Citizen group demands for access to information in the hands of government have also posed special difficulties, particularly in those countries where so-called "freedom of information" legislation does not exist. And information provided by government has often been criticised as not being politically relevant to the decisions at hand.

113. Information, if it is to have an influence in processes of public participation, needs to be provided on a timely basis, especially when there is a legitimate choice among options, and should be shared equally

by all parties. Citizens, themselves, should be able to have a say as to what information they feel is politically relevant to decision-making and what issues they feel deserve public discussion and debate. Public access to industrial information presents a crucial problem, especially in those cases where only one industrial proponent is involved. Special attention needs to be given to identifying under what circumstances and conditions such information should be made publicly available, while at the same time ensuring industry's legitimate claim to protection of proprietary information.

114. Similarly, with respect to information programmes designed to promote public understanding on complex technological and environment-related matters, non-governmental institutions can and perhaps should, in many circumstances play a leading role. Efforts should be made to ensure the growth and development of more pluralistic sources of information on environmental matters and, through education reforms, to increase citizen awareness and understanding of environmental issues and ecological principles.

115. The use of expertise and counter-expertise has become an important element in many recent environmental controversies, especially those where risks and uncertainty are difficult to measure. The availability of alternative or counter-expertise is instrumental as a means for encouraging increased public environmental awareness and for facilitating the tasks of policy-makers and parliamentarians. Without it, politicians can become captive of the government bureaucracy and its selected expertise. Providing for counter-expertise can contribute significantly to the maintenance of democratic processes and serve as a check against what some people see as technocratic tendencies in government. And yet, environmental expertise still tends to be concentrated and, in mots countries, is not widespread.

116. Efforts need to be made to ensure broader citizen expertise in environmental matters. In some instances and on certain issues, this may require government financial assistance to citizen groups for the purpose of developing their own expertise. At the same time, efforts should be made both to ensure that different forms and types of expertise are brought to bear on decisions and that experts' involvement in policy debates does not pre-empt political choices. For, one of the challenges confronting government to-day - in environmental and other policy-related matters - is that of integrating scientific and technical facts with political and social values. There are, of course, no easy, simple or proven methods for resolving this dilemma. Much research, development and testing, especially for the assessment of environment risks, is needed. At the same time, processes of participation can help to clarify, and hence to identify the types of choices on which political decisions are to be made.

117. Experiences with public participation indicate a general polarisation within the general public; a few citizens are often actively involved while a majority remain passive. This can lead to a form of participatory "elitism" in which those with the talent, know-how and energy may become disproportionately more "powerful" than the groups they represent.

118. This suggests the need to ensure that those citizens who do not desire to be actively involved in public policy debates and decision-making nevertheless retain their right to express directly their views and opinions. At the same time, care should be taken in defining under what circumstances "representativeness" should serve as an important criterion for the granting of public access to decision-making forums.

119. The need to develop closer linkages between public participation and government decision-making has been demonstrated at national, regional as well as local levels. For many people, participation without the possibility of influencing decisions is considered to be too limited. And yet, clearly, direct public participation in the exercise of decision-making power is not always feasible.

120. Specific measures should be taken to bridge the distance between decision-makers and the general public, and to identify under what circumstances government delegation of decision-making power to citizens is practicable. At the same time, consideration should be given to identifying the conditions under which it is desirable to forego "timely" and "efficient" decision-making in order to ensure appropriate processes of public consultation and consensus-building, even if such processes are more lengthy and more time-consuming.

121. Finally, there is a strong indication that demands for public participation in environment-related government decision-making will remain and perhaps increase. Over the last decade, it has already led to transformations of some governmental structures and processes. It has contributed to a changed outlook on the part of government and in the way it perceives citizen desires and needs. It has influenced directly public attitudes toward government decision-making and led to increased willingness on the part of many citizens to be more directly involved in environment-related policy debates. Public participation, then, can be seen as an essential means for increasing environmental as well as political awareness, and for seeking social consensus on the balance to be sought between economic development en environmental concerns.